The Mythology of Venus

Ancient Calendars and Archaeoastronomy

Edited by Helen Benigni

With a foreword by Morgan Llywelyn

University Press of America,® Inc.
Lanham • Boulder • New York • Toronto • Plymouth, UK

Copyright © 2013 by University Press of America,® Inc.
4501 Forbes Boulevard, Suite 200, Lanham, Maryland 20706
UPA Aquisitions Department (301) 459-3366

10 Thornbury Road, Plymouth PL6 7PP, United Kingdom

British Library Cataloguing in Publication Information Available

Library of Congress Control Number: 2012952329
ISBN: 978-0-7618-6062-4 (paperback : alk. paper)—ISBN: 978-0-7618-6063-1 (electronic)

Contents

Foreword

Morgan Llywelyn

Life is a school. There are few limits to what a human being with an open mind may learn. An open mind allows one to cross borders and transcend limitations. Such interdisciplinary studies challenge the flexibility — the range — of the human intellect. In this groundbreaking work the authors challenge us to look beyond the galaxy of our prejudices to a universe of possibilities.

We are flesh and spirit. Both are necessary to the whole. Separating the human body from its animating force leaves but a handful of minerals unable to reform into a sentient being. Modern science, far-reaching and complex as it is, cannot create a flesh and blood human brain with its unique individuality and awesome potential for creativity. That brain and that potential define us. We are no closer to understanding how and why than we were when we first stood on our hind legs and walked out of Africa.

Science asks us to accept only that which conforms to current mathematics or can be experienced through our five senses. Yet dogs recognize thousands of scents we cannot even smell. The common housefly sees a vast range of colours far beyond the human spectrum. They and the creatures like them inhabit a world invisible to us. Does it exist? Simple logic tells us it must.

Faith has been described as the evidence of things unseen. In an attempt to reify the non-corporeal, the world's leading religions have used priests, or prophets, or scriptures, or images, or even architecture. None of these explain the deeply ingrained human need to *worship*. It appears to be genetically implanted in our species, no matter the race or colour. We may worship at the shrine of Christ or Buddha; Elvis Presley or Marilyn Monroe. But in every generation, we worship. The need to direct our spiritual passion towards a receptive force is as powerful as our need to breathe.

This book examines some related aspects of the human need to worship. From the work of highly respected scholars in different fields it is possible to gain an overview of a unifying ancient symbol which is more than the sum of its parts. Since prehistory, mankind has recognized and responded to the sacral element in the female. No single image is sufficient to embody something which transcends time and space. The Morrigan, Celtic goddess of war, is Aphrodite, Greek goddess of love. Her terrible, tender face in a hundred different guises peers out at us from our collective memory. Embodiment of the life force, hers is the name men cry aloud in their death throes.

She is The Goddess.
    Morgan Llywelyn 2011

Preface

Before being introduced to the wealth of details and facts that lead to the overwhelming question of whether the archetype of the Goddess of Venus began in the Paleolithic cave art and ritual buried deep in the consciousness of our humanity, I would ask the reader to examine the motives put before this writer to inspire me to the task of this meticulous research project. What would make me pursue the project with an aim toward careful proof that such an archetype might be traced to the beginnings of religious and sacred thought and still influence us today? The answer is two-fold in nature. Primarily, I sought to prove, with the facts before me, such as those symbols engraved in the stone monuments of the distant past, that in the study of an archetype of a goddess of such magnitude, I would discover an enduring feminine presence embracing cultures over eons of time. In a world where the study of the archetype of the Goddess of Venus has been pushed aside as a representative of sexual pleasure and entertainment alone, the underlying importance of a female deity responsible for the resurrection and renewal of life has been understated if not buried in time. The feminine archetype of the Goddess of Venus as a symbol of resurrection and a belief in the afterlife is an essential element of the human psyche. Therefore, the full potential and powerful influence of the Goddess of Venus may be realized within the scope of this study which attempts to bring her characteristics, images, and symbols to life. This, in itself, is worth proving the fact that in our consciousness, humanity needs this female symbol for renewal both in a cultural and spiritual sense.

Second, I would put forward the idea that the study was conducted as proof that we are indeed the sum total of our ancestors. As Joseph Campbell has stated in volume after volume of intensive study of the archetypes of all cultures, the need to identify with the past and continue a mythology of our culture is also an inherent need in the human psyche. In order to accomplish this tremendous endeavor, we must study where we have been to continue the process of the evolution of human thought. At a time when several cultures have recognized the beginning of a cycle ripe for cultural change which is logically based on the world around us and the cycles of precession in the night sky that may very well communicate the need for change on some unconscious level, it is time to consider which myths are most abiding and which myths we have cast aside in the past that should be resurrected into a new, vital mythology. Entering the archetype of a goddess, such as the one associated with the planet of Venus, is one necessary part of resurrecting those images of the feminine that have been misunderstood or omitted in the patriarchal cultures that dominate the world today. However, the larger task before us is to re-create an entire mythology based, not only on the resurrection of the Goddess of Venus, but of a re-vamping of human thought conscious of the balance of the life force itself with a respect not only of a feminine and masculine balance, but one which contains a balance with the forces of Nature and the Earth itself.

To this end, we gather our thoughts around the ancient calendars of the Mesoamerican, the Greek and the Celtic cultures that introduce us to the startling date of the Winter Solstice of 2012 which might be one of the largest cycles of precession measured by humanity. Whether this precessional cycle is based on astronomy and/or religious speculation, the fact remains that we have accumulated an amazing amount of energy and human thought on such an avenue for possible change. It speaks of the need for change on a level of global consciousness never before presented to us as a species. Might we not pay some attention to the need for incorporating change into our mythologies which are expressions of the way we interpret and act upon the world around us? After having been involved in a decade of the study of the cycles of precession on ancient calendars, I would hope that the presentation of the archetype of the Goddess of Venus would be an inspiration for that change, for surely, it is the culmination of that study. The true balance of the observations of the celestial bodies that guide us through our own mythological life cycles are composed of three guiding forces: the sun, the moon, and Venus. It is this trinity of celestial bodies observable to the naked eye for both the ancients and ourselves that makes a calendar which keeps time for humanity thus directing us to be in tune with the larger cycles of our lives. And it is the cycles of Venus that complete that calendar.

The Mythology of Venus: Ancient Calendars and Archaeoastronomy is the third and final text in a series of studies that reveal the underlying myths of the cycles of precession. The first study, *The Myth of the Year: Returning to the Origin of the Druid Calendar* (University Press of America, 2003), demonstrates how the ancients based their myths on four seasonal groupings of constellations that follow the yearly cycles of the sun and the moon. For the ancient Celts and Greeks that involved keeping time by the night sky as it followed the agricultural cycles of the Earth. The goddesses and gods of the Earth, such as Demeter and Dionysus, lead the people through myths that keep humanity in tune with the movements of the celestial bodies. The second study, *The Goddess and the Bull: A Study in Minoan-Mycenaean Mythology* (University Press of America, 2007), focuses on the cycles of the moon beyond the year. Here, the five, nineteen, fifty-six, and other successive precessional cycles of the moon have been unearthed in the Bronze Age culture of the Minoan-Mycenaean peoples as well as the Celts to form a body of religious ritual and belief whose central figure is The Goddess and Her consort, The Bull. This archetype forms the basis of understanding the concept of sacrifice at the center of many mythologies that follow this ancient example. Finally, we have come to the cycles of Venus which must be intertwined with the precessional cycles of the sun and the moon to form a trinity as the basis of many belief systems of ancient cultures. Venus emerges as the Goddess of Resurrection and Renewal who gracefully moves through the celestial patterns climbing the heights of the night sky as she rises from the sea in her own cycle that transgresses the yearly cycle flowing in and out of the consciousness of both lunar and solar energy.

In the first chapter of *The Mythology of Venus* entitled "The Emergence of the Goddess," I identify the archetype of the Goddess of Venus in the Paleolithic cave art, sculpture, and ancient monuments in its earliest stages. Venus appears to be part of what Carl Gustave Jung and Erich Neumann call the transformative character of the primordial archetype where her image is seen as a regenerative force for change connected to the celestial order. Joseph Campbell and Marija Gimbutas also note the emergence of the transformative form of the archetype as time-factored and identified with cyclical time in the cosmos, respectively. Lascaux Cave and Pierre Plates on the Locmariaquer peninsula in France use symbols such as sacred water, water birds, reclining nudes, and the columns of life to delineate the patterns of the planet Venus in its cycles as Morning and Evening Star. The fluidity and beauty of these patterns in the night sky represent a feminine force for conception from the maternal waters as well as the regenerative force of life itself. Like the caves in France, such as Lascaux and Chauvet, the Neolithic mounds of Newgrange at the Brú na Bóinne in Ireland and the Neolithic temples, such as Ħaġar Qim and Ħal-Saflieni in Malta, also embody the concept of Venus as Regeneratrix in both the construction of their sacred spaces and in symbolic language carved on the temples. The sacred waters, reclining nudes, triangle signs, and the figure of a goddess climbing The World Tree are dominant representations in the cycles of Venus reflected in the temples' orientation to the planet's journey in the night sky. The most intricate patterns of the cycle of Venus are seen in the notations of astronomy on the kerbstones of Newgrange carved by the ancient astronomers of Ireland.

The cosmology of the sacred feminine represented by the Goddess of Venus develops in the Minoan-Mycenaean culture of the Mediterranean in what is later depicted as a goddess with arms raised in a posture of adoration of the heavens which is represented in the Greek alphabet as "psi." This goddess embodies the concept of spiritual enlightenment through awareness of the celestial and as a model for transformation is identified with many of the same symbols of renewal such as sacred water, water birds, the columns of life or The World Tree and beautiful reclining nudes. A *kouros*, or young male companion, is depicted in the frescoes and sculpture of the Minoan-Mycenaean culture as a companion that the Goddess of Venus must retrieve from the Underworld in her journey through the night sky. Lustral basins, pillar crypts, myrtle trees, doves and other iconography of the Goddess of Venus are seen in the expansive cult centers of Knossos and Mycenae where the goddess' temples are aligned to the planet's cycles. In a chapter entitled "The Epiphany of the Goddess," I identify these images as precursors to the figure of Aphrodite in Greek mythology. An easy transference of the Goddess of Venus as Aphrodite and her lover as Adonis in Greek mythology follows with the explanation of the myths that trace the many faces of Aphrodite as Morning and Evening Star. Her identification with the powers of creation, her birth from the sea, and her journey to eternally re-unite with her lover corresponds with the cycles of the planet Venus. Likewise, Aphrodite's temples in Greece, such as those located on the Acrocorinth and the Acropolis, are aligned to Venus' journey in the heavens. Many festivals on the Attic Calendar, including the Adonia, the Skiraphoria, and the Aphrodisia, re-create the myths of the Goddess of Venus using rituals and symbolic art to enact the journey of the goddess and planet.

The cultural importance of the Goddess of Venus is broadened by Miriam Robbins Dexter's chapter that cites the images, myths, symbols and art associated with the goddess in Indo-European cultures. In a chapter entitled "Love Goddesses of the Early Historic Age," Dexter identifies female figures associated with love goddesses in the Paleolithic and Neolithic Eras to include figures with apotropaic features, or those goddesses who protect the worshippers in the temple or cave, as well as figures associated with musical instruments, dance, and sacred displays of genitalia. According to Dexter, "the female body was considered sacred and propitious, and the nudedancer was thus able to effect the magic needed for the group" as a shamanic figure. Indus Valley figures as well as figures from Anatolia are identified with the love goddess as Venus by waterbirds and snakes as part of their iconography of symbols that represent the life continuum and regeneration. Dexter cites the early historic goddesses that repeat these iconographic features such as Inanna in ancient Sumer. As Inanna and Ishtar among the Akkadians and

Babylonians, she is also the Queen of Heaven, the bestower of love and fertility who searches the Underworld for her lover, an attribute clearly associated with the journey of Venus in the night sky. In Egypt, one function of Hathor exaulted in poetry, music and dance is as love goddess, and a similar function is associated with Shrī Lakshmī, "born of the foam at the churning ocean," Durgā, and Tārā in the Hindu texts. Dexter concludes her study by identifying the roots and rituals of Aphrodite in the Classical World as well as the etymological roots of the planet Venus as a wanderer in search of her lover.

What becomes evident in the comparative mythology of Venus in several ancient cultures is the reoccurring story of a goddess whose role is one involved with the patterns of the planet in the night sky. Anthony Murphy in a chapter entitled "Venus, The Caillichín na Mochóirighe of Newgrange, Ireland," takes a look at this role in terms of the mythology and archaeoastronomy in the Boyne Valley monuments of Ireland. Murphy carefully outlines the overall purpose of the monuments in the Boyne Valley, which includes the mounds of Newgrange, Dowth and Knowth, and their cultural and mythological significance. The patterns of Venus are part of an ancient cultural myth that explains the Winter Solstice as a time of birth and re-birth. The Caillichín, or virgin goddess of the Boyne who is associated with a sacred white cow and white swans, rises once every eight years before the sun as the Morning Star whose light fills the passage and chamber of the mound of Newgrange followed by the light of the sun which has the same effect. This celestial phenomena is recounted in a series of myths about the Boyne, including the tale of Bóann, the *cailleach* goddess for whom the river Boyne was named; the tale of Diarmaid and Gráinne and the birth of their son, Aonghus who is born in one miraculous day; and the tale of Cúchulainn and his birth. Stories of a miraculous virgin birth from the sacred mound associated with the goddess who gives birth, as well as the appearance of the last crescent moon beside the Morning Star, represent the cycle of birth and rebirth where myth is part of the eternal repetition of the celestial bodies in the night sky. When the meeting of Venus, the sun, and finally, the crescent moon occurs in these tales, the ancient calendar is complete.

The mythology of Venus incorporates art, archaeology, and astronomy into a coherent form to elucidate the phenomena of the journey of the planet in the night sky. As the brightest of the planets and the brightest object in the sky except for the sun and the moon, Venus travels through the morning and evening sometimes rising higher than the sun, and then descending reaching its furthest place on either horizon. For many cultures from the beginning of time until the current era, Venus' journey has many meanings most of which involve a divine being, a goddess, imparting the renewal and resurrection of the life forces through the belief in love. Venus, sometimes called The Queen of the Heavens, is seen first after the sun sets, and then all of her subjects, the stars, appear to decorate the night sky with immortal and divine light. Without this traveling celestial body, the regeneration of the cycles of the universe seem impossible and without cause. Surely, it would seem to the ancients and to our own modern eyes that we are subject to a motion that is not our own.

Helen Benigni
April 13, 2012

Acknowledgments

Eadhmonn Ua Cuinn, Barbara Carter and I began to translate the Coligny Calendar in 1999 and have never really stopped conversing on the subject of ancient knowledge. Where Eadhmonn recalls Celtic cultural knowledge from a deeply rooted sense of understanding and perhaps an intuitive, artistic knowledge from his work as a stone carver, and Barbara uses an ethnographic approach to deciphering codes of astronomy, I am able to interpret the symbolic language of ancient myth and trace its roots from prehistoric times. For this gift of insight that has allowed me to complete a study of astronomy and mythology, I acknowledge these people.

Moreover, in the study of Venus and the mythology associated with the cycles of astronomy, I would like to acknowledge the gift of esteemed scholars who have broadened the scope of my research. Morgan Llywelyn has given her moral and written support over the years. Miriam Robbins Dexter has become a dear intellectual companion and friend as well as the only person I have ever truly trusted to edit my work, and Anthony Murphy and R.P. Hale have entered our circle to give expertise in their respective fields. The coming together of such expertise and friendship is rare and should be cherished.

Finally, of course, I would like to thank family and friends in Elkins, and colleagues and students at Davis and Elkins College, for their continued support and advice.

Helen Benigni, 2012

Chapter One

The Emergence of the Goddess

A Study of Venus in the Paleolithic and Neolithic Era

Helen Benigni

In the evolution of human thought, both Carl Gustave Jung and Erich Neumann mark the development of what they believe are primordial archetypes. Neumann explains that the primordial archetype is most clearly defined in the writings of Jung as "the structural concept signifying 'eternal presence'" manifested early in the stage of human consciousness before differentiation into particular archetypes. This myriad faceted stage of the archetype leads to the emergence of individual archetypes from a great complex mass to the formation of coherent archetypal groups. Parallel to the development of the primordial archetype into individual archetypes is the creation of symbol, symbol sets, and conscious ritual. What follows, eventually in the origins and development of human consciousness, is the organization of those symbols into mythology. For example, Neumann refers to the primordial archetype of what he calls "the way" or path of individual discovery beginning in the exploration and establishment of caves as temples adorned with magical and sacral art to the conscious ritual associated with shrines and temples in early myth (*The Great Mother* 7-9).

The formation of the primordial archetype of the sacred feminine follows similar lines of development merging into what Neumann terms "symbolic polyvalence" diversity, and multifarious diversity containing contradictory elements (*The Great Mother* 9). Accordingly, the primordial archetype of the goddess begins in the Upper Paleolithic era with the appearance of Venus figurines, the oldest to date being the Venus figurine recently discovered in Hohle Fels, Germany dating approximately 35,000 years ago. As many as 400 figurines depicting similar characteristics have been found in Central Europe and the Mediterranean as well as a few in sites in what was to

1

become Anatolia or what we know as modern Turkey. The amazing similar-
ities in figure, shape and sacral context reveals evidence as to the existence of
a religious sentiment expressed in these figures. Although quite far from the
establishment of organized religion, these Venus figurines are indicative of
what Jung and Neumann see as the primordial or primeval, non-derivative
expression of the idea of the feminine as sacred.

In the development of the primordial archetype of the feminine, variations
or characters although sometimes multi-faceted and diverse may co-exist.
Neumann distinguishes two distinct characters of the Venus figurines of the
primordial archetype: the elementary and the transformative characters (*The
Great Mother* 24-6). The elementary character of the goddess displays itself
in what Neumann calls the Great Round or Great Container that "tends to
hold fast to everything that springs from it and to surround it like an eternal
substance," whereas the transformative character of the primordial archetype
expresses the need to change and is based on the blood-mysteries of the
feminine. Neumann states that "the two characters are not antithetical from
the very start but interpenetrate and combine with one another in many ways,
and it is only in unusual and extreme constellations that we find one or the
other character isolated. But although both are usually present at once, one of
them is almost always dominant" (29). For example, in a Venus relief carving like the Venus of Laussel what Neumann states is clearly obvious in the
sense that the goddess' shape with long pendant breasts and massive hips
delineate her purpose as container of life exemplar of her elemental character
while the crescent-shaped auroch horn that she holds delineates her dominant
purpose as keeper of female cycles and lunar time. When referring to the
Venus of Laussel, Joseph Campbell notes the emergence of both characters
of the archetype of the feminine as "time-factored." Campbell states that: "It
may therefore be that the initial observation which gave birth in the mind of
man to a mythology of one mystery informing earthly and celestial things
was the recognition of an accord between these two 'time-factored' orders:
the celestial order of the waxing moon and the earthly order of the womb"
(*The Way of the Animal Powers* 68).

While both characters, elementary and transformative, of the primordial
archetype are equally necessary, the dominant mode of the depiction of the
goddess outlines a progression that is followed through the Paleolithic art
into the Neolithic, Bronze and Iron Ages that acts as the basis for the formation of the archetypes of goddesses in several cultures. Hence the characters
of the primordial archetype lay the foundations for the goddesses of religions
in the development of consciousness and culture. According to Neumann, the
study of the transformative, regenerative character of the archetype connects
with the celestial order of the universe as Regeneratrix, an intermediary
position between worlds establishing the lineage of the goddess of regenera-

tion, and cyclical and cosmic time whereas the goddess of earthly concerns has roots in the elementary character of the primordial archetype.

Marija Gimbutas makes a statement concerning the formation of the archetype in her studies by establishing the sacred feminine and its characteristics. In Marija Gimbutas' text *The Language of the Goddess* (New York: Thames and Hudson, 1989), Gimbutas defines the emergence of the goddess and the symbolism attributed to the goddess as containing the mystery of birth and death and the renewal of life as well as its relationship to the cosmos. Gimbutas states: "Symbols and images cluster around the parthenogenic (self-generating) Goddess and her basic functions as Giver of Life, Wielder of Death, and not less importantly, as Regeneratrix, and around the Earth Mother, the Fertility Goddess young and old, rising and dying with plant life"(xix). Gimbutas adds that the goddess is the single source that took the energy from the sun and the moon and the life on earth in a cosmos that was represented by cyclical and not linear time. Gimbutas' referral to the connection between the formation of the archetype and its association with the celestial cycles, like Campbell's reference to the connection between cosmic time and the archetype, emphasizes a signifier of the primordial archetype whose development leads to the emergence of a goddess of the celestial cycles of the sun, the moon, the visible planets, and the constellations in Neolithic, Bronze and Iron Age cultures.

Although the definitions of the elementary and transformative characters of the primordial archetype depict a co-existing duality of forms, and although Neumann, Campbell and Gimbutas delineate the earth-centered attributes as well as the regenerative and cosmic-centered attributes, the existence of a triad or trinity of feminine images also begins in the Paleolithic era and is duly noted in the emergence of the primordial archetype. This triad of feminine figures seen as the mothers of the Rhineland at Gönnersdorf from the Magdalénien period, the three inverse figures in low relief at Laussel, and the cluster of claviforms, or full-bodied female images, above the entrance of Lascaux Shaft all testify to a multiple image of the primordial archetype. This seeming contradiction establishes the importance of a triad of power emanating from the sacred figures that continues throughout the development of the primordial archetype into the formation of archetypes of several cultures in the Neolithic, Bronze and Iron Ages. Therefore, the complexity of the primordial archetype contains both elementary and transformative characters, both orientation to earth and sky in its relation to the feminine mysteries, and is expressed in the power of the triad. Moreover, its complexity is magnified by the merging of all these characteristics often with one more dominant from the others, but with all magnificently present. This forms what Neumann so aptly calls the essential representation of the complexity of the feminine and what Campbell most affectionately admires in the Venus of

Laussel "that stands before us like the figment of a dream, of which we dimly know but cannot bring to mind the meaning" (66).

Without losing the idea of the complexity of Paleolithic art and the depiction of the primordial archetype in its original form, the emergence of the Regeneratrix, the goddesses' connection to regeneration, rejuvenation, and cosmic and/or cyclical time, may be traced through the ages in order to signify the importance and re-establish the meaning of the archetype in mythology. The transformative character of the Regeneratrix has a cluster of images that are distinguished by two paths, both of which are oriented around celestial imagery. The first path of concern is the depiction of the Regeneratrix in connection with the moon and the feminine blood-mysteries, and the second path is connected with mutable change and the patterns of flux and change that are part of a more flexible pattern than that of the moon. The first pattern of the Regeneratrix's time-keeping of the lunar cycles is represented by the auroch's horn of the Venus of Laussel, crescent moons, the shape of the "U" to represent the cycles of lunar precession beyond the yearly cycles, and the dots of red and counting marks found in Lascaux which most likely represent the cycles of menstruation and pregnancy. The second pattern is represented by fluid marks of water, sacred water birds, emerging and reclining nude females, and an association with the columns of life.

Although precise notations of the cycles of the planet Venus do not appear until markings on the megalithic stones of the Neolithic era, such as those noted on Pierres Plates on the Locmariaquer peninsula in France and on the Brú na Bóinne in Ireland, the distinguishing characteristics of the Regeneratrix in the Paleolithic era most closely identified with the cycles of Venus are those beginning with the sacred water and water birds, the reclining nudes, and the columns of life. These patterns, found on sculpture, cave drawings and on natural cave structures, have a mythological theme that pertains to the beginnings of life from uterine moisture, the miracle of conception, and the emergence of life from the waters of the womb. The path of the planet Venus represents these mysteries because its path, quite visibly noticeable to the ancients, not only unites the patterns of the moon and sun into a feasible calendar and time-keeper that is easy to follow and used in all ancient calendar systems including the Greek, Mayan, and Celtic calendars, but Venus also re-enacts the cycle of conception to birth in her rising as the Morning and Evening Star like a bird emerging from the watery realms. The periods of visibility of Venus between first appearing as Morning and Evening Star and vanishing takes an average of 260 nights that correspond to the time between conception and birth. Hence, Venus becomes an apt representation of the emerging forces of life from the waters of conception.

Furthermore, the images of the Regeneratrix associated with conception also connect the belief of a celestial watery realm to the body of women, often seen as water birds, and the mysteries of transcending one spiritual realm to another. Unlike the patterns of regeneration evident in the symbolism associated with the lunar cycles, the patterns of Venus ascending and descending from water to air are clearly marked by the five positions of Venus in the night sky. To the ancients this is represented by movement from the water to heights in the night sky. Markings on the columns of life in the Paleolithic stalactites are later transferred to the images associated with The Tree of Life and the goddesses' ladder in the Neolithic era. The symbol of the pentagram also has its origin in the orbit of the planet of Venus. Finally, because Venus appears to travel through the sky close to the sun, it is also associated with brightness setting itself apart by its exceptional beauty and blue-white light and its ability to reflect light from the sun. These attributes are associated with the beauty and sexual appeal of women represented in the reclining nudes at La Madeleine cave and the figurines of Brassempouy in the Paleolithic art through time to the figures of Aphrodite in Greek mythology. The planet's visibility and the dense cloud atmosphere surrounding the planet result in certain phases where Venus will outshine most objects in the night sky. Such magnificence at sunrise or sunset combined with the changing patterns and rising from the sea were noted by the ancient astronomers as miracles of life fit to be worshipped as sacred.

Even the disappearance of Venus below the horizon for 65 nights of its 584 night cycle is connected to fertility rituals and a return from the underworld or sleep to begin a new cycle through possible conception and birth. The largest cycles of Venus, a return to the same place in the night sky every eight years, also marks a time for celebration because the planet, seen as the Regeneratrix, coincides with 99 moons, seen as primordial goddess most closely associated with the moon. What began in the Paleolithic caves as an ideology and representation of the transformative or Regeneratrix aspect of the sacred feminine gains momentum and distinct expression in the evolution and emergence of the sacred feminine through time. The symbol of the ladder, represented in the night sky as the path of Venus, is described by Mircea Eliade as an important archetype in primordial symbolism as such: "It gives plastic expression to the break through the planes necessitated by the passage from one mode to another, by placing us at the cosmological point where communication between Heaven, Earth and Hell becomes possible" (*Myths, Rites, Symbols* 239). Thus, the planet and the goddesses it represents are transformative and regenerative in nature based on natural phenomena and the objective correlative of accurate observation and experience. We are connected to The Axis of the World and the cosmic mysteries of conception through the belief in the beauty and mystery of the feminine.

The fundamental aspect of the Regeneratrix as Venus is what Gimbutas calls the goddess as "Mistress of the Upper Waters" where the sacred feminine is worshipped in the Upper Paleolithic in the form of water as a life-giving source connected to the cosmos. Gimbutas states that at the Magdal-

enian sites of Montespan and Tuc d'Audoubert in France, a stream flows
from the mouth of a cave (*The Language of the Goddess* 43). Likewise, at the
site of Les Eyzies-de- Tayac where underground water flows from the sacred
cave of the goddess through the town, the worship of water is evident in
connection with the goddess. Many Paleolithic sanctuaries in caves contain
lakes and subterranean rivers associated with what Gimbutas calls "a long-
lasting belief in the magical potency of streams and wells." On the Upper
Paleolithic and Mesolithic figurines Gimbutas notes: "Streams in the form of
'comets' or parallel lines in diagonal, vertical, or meander bands, are frequent motifs of Upper Paleolithic and Mesolithic mobiliary and parietal art."
These comet streams are found on female images symbolizing the "divine
moisture with the body of the Goddess" (43-4). On the Roc de Marcamps,
Gironde, France, the comet striations closely resemble the pattern of Venus
in the night sky where the planet moves from the horizon as the Eastern Star
to the Western Star and then under the horizon forming a definite cone shape
movement in later mythologies identified with the goddess climbing The
World Tree.

The pattern of the movement of the goddess from the underworld to
traverse the cosmos bringing the moisture from below into a transmogrifying
blue comet-like striation of light and life is also marked on the bodies of the
goddess in Paleolithic art where the goddess resembles a water bird. On a
plaquette from Gönnersdorf, Germany the bodies of three water-bird goddesses are striated with net patterns and filled with divine moisture. Likewise, a relief of a bird goddess at Pech-Merle, France from the Early Magdalenian is over-painted with water striations as are many bird-goddess figurines
from the Paleolithic era. Gimbutas comments: "The Bird Goddess was the
Source and Dispenser of life-giving moisture, an early and enduring human
preoccupation. As a waterbird, she united heaven and earth, and her terrestri-
al home was probably believed to be mirrored by a celestial watery realm"
(29). In later mythologies, the association of the water bird with the goddesses of the Neolithic, Bronze and Iron Age cultures persists most evident in the
association of Aphrodite with the swan and the goose and with Caer and the
swans at Newgrange, Ireland.

As Mistress of the Upper Waters, the bird goddess' flight represents a
type of epiphany where the feminine source of birth and creation traverses
realms to reach celestial or cosmic heights. Her journey follows the patterns
of Venus in the night sky and her image as a water bird emerges clearly in the
Paleolithic era as an important element. Moreover, there is evidence that
rituals in the cave sanctuaries used the naturally formed stalactites at the back
of caves to represent The Column of Life, The World Tree, or The World
Axis that the Regeneratrix climbs. To demonstrate the journey of the Regeneratrix from the waters of creation symbolized by the subterranean waters of
the cave, the stalactites that formed a column were carved with symbols of

the goddess. Her emergence from the caves symbolizes the emergence of the planet Venus from below the horizon at the beginning of her cycle through the night sky. Much like conception in the womb, the rituals of the goddess in the caves are steeped in sexual metaphor and meaning. It is here where fertility rites most likely took place.

In the cave of Lascaux in the town of Montignac, France numerous female signs are present in the seven naturally formed chambers of the cave that lead to the sanctuary at the cave's end. The end chamber, called the Chamber of Felines, is reached by passing through several chambers where animals seem to run toward the end chamber aided by a male shaman. About sixty feet in the passageway from the first room called the Rotunda, are two rooms called the Apse and its Shaft. The formation of these rooms forms a 120° angle with the Apse as an ante-chamber to the angle. This exact angle formation and apse was built in the Neolithic structure of Pierres Plates in Morbihan, France near Carnac. In Pierres Plates, the tilted passage, the apse, the signs and symbols of the Venus cycle on the stones, and the location of the temple itself on the Bay of Morbihan attest to the fact that it too was aligned to the cycles of Venus and possibly modeled after Lascaux or other temple caves of the Paleolithic. In both Lascaux and Pierres Plates, the angled chamber serves as a view room where the light of Venus may be seen in the Shaft or Venus may be viewed from the Shaft. The Shaft has been noted by archeologists as a shaft that was definitely not used as an entrance.

The Apse is a small chamber of about five yards in diameter that is covered with thousands of engraved lines which are exactly like the comet streams on the female images symbolizing the divine moisture of the goddess. Here, too, are the only representations of the full-bodied female or what André Leroi-Gourhan identifies as the "claviform" of the goddess yet known in the Périgord (513). These female symbols are accompanied by quadrangular signs similar to the quadrangular signs on a necklace of what was most likely the high priestess' body discovered in La Madeleine ou du Roc-de-Cave in nearby Tursac, Dordogne dating 15,780 B.C. and exhibited in the Musee de Les Ezyies-de-Tayac. Many similar quadrangles used for counting lunar cycles of months and precessional lunar cycles are also seen in Les Eyzies. This out of the ordinary Apse chamber and Shaft comfortably fits two people standing or sitting, with a backdrop of a Paleolithic scene framed by an immense chevron or "V" sign partially composed from the natural formation of the rock and man-made carving. On the stone rim of the Shaft are several more engraved claviforms such as those at Le Tuc d'Audoubert. In the painting scene is a rhinoceros with six dots in two rows of three under his tail, barbed signs with lateral dots, and a figure of a bird, another symbol of the Mistress of the Upper Waters, as well as a speared bison, an ithyphallic man and several horses.

Although an exact interpretation of the meaning of the prehistoric animals in their mythological context is speculative, the location, size and unmistakable immense "V" sign as well as the other symbols and signs connected with the Regeneratrix and Mistress of the Upper Waters is clear. Birds are rare in cave art and the presence of the bird, the "V" signs, the claviform signs, the quadrangles, and, most importantly the shape and orientation of the Apse and Shaft may bring to light another view of this ancient sanctuary and passage temple. The most curious display in the Apse painting is the series of horses that are running toward the immense "V." The last horse in the series reaches the point of the "V" and is turned on its side. If the men and animals in Paleolithic art symbolize the masculine force entering the feminine sacred, which is the cave as the womb and/or uterus, then perhaps reaching the epiphany of the experience inside the womb and at the end of the Shaft at the point of the "V" might represent the orgasmic experience of losing one's self to the eternal female as sacred thus implying that the sex act itself is one of a sacred nature. In the monuments of the Neolithic, Bronze and Iron Age, the idea of sex as a sacred act in the name of Aphrodite, as well as parallel forms of this goddess of regeneration, are an essential metaphor in the worship of Venus found in Lascaux perhaps as one of the primeval temples of its kind.

Another cave temple attributed to the worship the goddesses of regeneration is the Chauvet Cave in the Ardeche Region of southern France. The end chamber of the cave, known as the Sacristy, is marked at its entrance with three abstract vulvas, and on its right wall is a painting called the Lion Panel. Opposite the Lion Panel, and the first image seen when entering the chamber, is a large stalactite with the painting of a half-man, half-bison seeming to mount a large vulva near the hip of the bottom half of a woman's body. The vulva depiction is a filled oval, pointing downward with a very concave base emphasized with a mark indicating the opening to the vulva. The shape of the stalactite is unmistakably phallic and fleshy in appearance, and the chamber is unmistakably shaped like the womb of the goddess. Judith Thurman, in a recent article in *The New Yorker* (June 23, 2008) remarks that the depiction appears to be part of the bottom half of a woman's body with heavy thighs and bent knees that taper at the ankle with a darkly shaded vulva and no feet, and the half-bison, half-man appears to have "an aroused eye." A line branching from his shoulder looks like a human arm with fingers, and the woman seems to be in a squatting position as if giving birth to the images of the bison and felines that line the walls of the chamber (64). Again, this chamber is a small, sacred place at the end of the cave that evokes a highly sexual metaphor and most probably was used for fertility rites.

The stalactite formation, the womb cave and the sexual metaphors in the Paleolithic are the primary images related to what Marija Gimbutas calls The Columns of Life. Like the image of the bird and the striation marks of water and life, The Columns of Life, later known as The World Axis or The World

Tree, is an inherent and necessary symbol in a complex of images associated with the goddess of regeneration and the planetary movements of Venus. The goddess in her many forms mounts The Columns of Life to her moment of epiphany and then descends into the underworld to conceive life from love in an ever returning cycle. Like the planet, rising and descending from the horizon in the east and the west, and then for a brief time remaining under the horizon, their paths are similar and cyclical. Gimbutas traces the origins of The Columns of Life from the Paleolithic caves of Dordogne and Tuc d'Audoubert to the Neolithic, womb-like caves such as those in Koutala, Crete and in Scaloria, Italy where more than 1500 vases with symbols of the goddess painted on them were discovered in the lower section of the cave (223). Gimbutas goes on to site caves used by the Minoans, such as the Cave of Eileithyia east of Herakelion and those at Psychro, Arkalokhori for rites associated with regeneration. On one Middle Minoan multi-faceted seal carved with a life column and a sprouting plant is "a whorl, or eight-pointed star," a symbol of the regenerative forces of Venus in its eight year cycle (222).

Numerous figurines embody the idea The Columns of Life as a vital force of regeneration by depicting what Gimbutas calls "a fusion of the phallus with the divine body of the Goddess, which begins in the Upper Paleolithic. Some of the 'Venuses' of this period have phallic heads and no facial features" (231). They have been found in northern Italy, Bavaria, France and, most importantly, in Cyprus, the mythical birthplace of Aphrodite in Greek mythology. On the figurines from Cyprus, Gimbutas notes: "Although the male element is attached, these figurines remain essentially female. They do not represent a fusion of two sexes but rather an enhancement of the female with the mysterious life force inherent in the phallus. The Goddess figurine creates a base from which the phallus, understood as a cosmic pillar, rises. It comes from her womb in the same way that stalagmites and stalactites grow from her womb in the cave" (232). In one "phallic figurine" from Cyprus, two views present anatomically correct images of both female and male genitalia where the female appears to be birthed from the testicles of the male when seen from the bottom. The figurine calls to mind the birth of Aphrodite herself. In Greek mythology, Aphrodite rose naked from the foam of the sea and stepped first on Cythera then on Cyprus, her principal seat of worship. According to Hesiod in *Theogony*, "As soon as he [Cronos] had cut off the members with flint and cast them from the land into the sea, they were swept away over the main a long time: and a white foam spread around them from the immortal flesh, and in it grew a maiden" (184-5).

Another characteristic of Venus seen in the emerging primordial archetype of the Paleolithic art that can easily be traced to later cultures, such as Greek and Roman culture, are the reclining, nude figures of the goddess. Although the furthest chamber of the caves with marked columns of life, such as those in the caves at Lascaux and Chauvet, are at the far rear of the cave, indicating a more likely place for fertility rituals, the entrance to the caves is where reliefs or carvings of female figures, such as the reclining nudes, have been found. At Les Eyzies-de-Tayac in Dordogne, there is a cave shelf with sacred water basins and an area large enough for a gathering, at Laussel, France, outside the cave entrance there are the reliefs of the triad of the goddess, and at Angles-sur-Anglin, Vienne three colossal female figures manifest above the head of a bull at the cave entrance. Most importantly, at the cave at La Madeleine, reclining nude women and large pubic triangles decorate the entrance. The most likely scenario in the ritual use of the Paleolithic caves would be that the entrance would serve for gatherings, and in the case of Les Eyzies and La Madeleine where the entrance is aligned northwest to southeast, they most likely were used for the siting of Venus rising as the Evening Star in the early or late spring while the inner chambers were used for fertility rites and rituals. In fact, the depictions on the outside or in the entrance area of the caves might act as an indicator as to the purpose of the caves.

At the entrance of La Madeleine, Tarn, which is some fifty miles south of Lascaux, on either side of the rocks are two reclining, nude figures that seem to lie upon the rocks in repose. H. Bessac, who discovered the female reliefs, remarks that, "'Both figures lie stretched out in positions of utter repose, one arm bent and supporting the head. They rise from the rock as "foam-born" Aphrodite arose from the sea'" (Campbell 69). According to a well-known archeologist, Leroi-Gourhan, the attitude of the women reflects "a nonchalant freedom" (Campbell 70). If these figurines evoke the images of Aphrodite in the Classical era of Greece and later in Rome where the goddess is identified by her reclining position or her nonchalant leaning position when standing, then these figures might well be precursors of the Regeneratrix as a beautiful, reposed woman. Classical statues of Aphrodite in the Agora of Athens are identified as Aphrodite because each figure leans nonchalantly against a post, a symbol of The Columns of Life or The World Tree; on the base of the Hephaisteion in the Agora, the goddess leans against a small column while cradling Eros at her side. Likewise, many of the Greek vases portraying Aphrodite's role in the marriage ceremony depict the goddess seated in a reclined position or leaning against a tree or column. The Roman copy of Aphrodite Olympias from the Circus of Maxentius seated in her reclining chair is perhaps the most impressive image of an alluring, relaxed goddess.

Another figurine that exists a bit closer in time to the reclining nudes of the Magdalenian era than those of Greece or Rome, is the figure of a reclining woman known as the Sleeping Lady of Malta from the Hypogeum of Ḥal-Saflieni. Discovered in a subterranean chamber with a four-columned trilithon altar, the Sleeping Lady and the two other goddess figurines found with her are examples of both the reclining attitude of the Regeneratrix and the early cave temples that were originally caves and then underground temples. In the side chamber, or apse, is an oracle room that produces a powerful acoustic resonance from any vocal sound made within it. The subterranean temple, like other temples of the Neolithic, Bronze and Iron Ages, includes the same elements of the goddess temples of the Paleolithic era. Primarily, the temples were all modeled after the sacred caves with their walls painted in red ochre to imitate the womb of the goddess as Regeneratrix. Second, an oracular element, like the area for speeches on the entrance shelf at Les Eyzies-de-Tayac or the oracular chambers at Ħaġar Qim, Malta and Delphi, Greece, serves as a medium for the voice of the goddess to reach the people. Additionally, each temple or cave has an altar for the goddess in the shape of an apse with claviform and triangle signs, or in the case of the temples, a trilithon often with a large stone in the shape of a pubic triangle. In the Malta temples, the apse is entered through a stone doorway with a circle in the center perhaps to imitate the cave entrance to the subterranean world.

The temple-complex at Ħaġar Qim is a megalithic temple-complex facing the sea on the island of Malta dating to approximately 3600 B.C. with a trilithon entrance, chambers of red orchre, and doorways that open to small chambers, one of which is an oracle chamber. Most importantly, Ħaġar Qim was most likely used as a temple of the Regeneratrix associated with the planet Venus. The temple-complex faces the sea on its western side giving full view of Venus rising from the sea as the Evening Star in three distinct positions in relation to the moon. Each of three chambers on the western side of the temple-complex has outside entrances that orient to these celestial sitings. A small semi-circular niche housing a triangular altar stone outside the temple faces northeast to site the solstices, equinoxes, and the rising constellations as well as Venus. On the east side of the temple the oracle chamber views Venus as the Morning Star in position from due east. The third siting of Venus in the east as the Morning Star is through the Main Entrance which faces southeast, the position of Venus furthest in the east, most likely associated with her beginning and/or ending of each complete cycle. The Main Entrance itself is a spectacular trilithon with eight entrance stones, the number of the full cycle of Venus in conjunction with 99 moons.

The approach to the temple is marked by an orthostat and several ruins of older phases of the temple as well as Il-Misqu or the watering place. The fact that Ħaġar Qim went through several stages of completion, each successively more complicated to represent the cycle of Venus more accurately, emphasizes the importance to the ancients of the Regeneratrix in her transformative state. The watering place itself also testifies to the importance of water in the cycles of earthly and celestial life. It is a flat area near the temple-complex that contains eight bell-shaped reservoirs that still retain water, eight again noting the Venus cycle. A monolith is near one well and a re-planted fig tree

near another, perhaps signifying the goddess in her water character and identifying her with The Columns of Life or The World Tree, respectively. The
fig tree is a symbol associated with the Regeneratrix in Minoan symbology
where the goddess is seen in full glory as she mounts The World Tree
depicted as a fig tree on the Ring of Minos worn by the high priestess of
Knossos, Crete. The orthostat due north of Ḥaġar Qim, the monolith near one
of the wells at the watering place, and several representations of The World
Tree or World Axis present inside the temple-complex of Ḥaġar Qim are a
reminder of the journey of the goddess in the night sky rising from the
underworld and traversing the celestial and watery realms. The World Tree
points to the pole star or due North, the center of numinous energy, much like
the orthostat of the goddess that guards the temple-complex of Ḥaġar Qim.

In a chamber to the left of the Main Entrance of Ḥaġar Qim is another
representation of The World Tree: a four-sided pillar. The four-sided pillar
has a tree and an owl on each side of its base, a large bowl-shaped basin for
water on its top, and cup-marks over it. All these carvings are images to
reinforce the importance of the pillar as the The World Tree and the association of the goddess with the waters of renewal and new life. Next to the pillar
is a carving of a decorated "V" with spirals on its legs and a bas relief of
plant life; these images also establish the idea of regeneration from a feminine source. Again, the image-complex of water represented in the bowl, the
bird with deep-set cup marked eyes, and the tree work together in the context
of a growing ideology or a language of the goddess in symbolic form. Even
the representation of twelve leaves on each side of the pillar might be connected with the twelve moons of the year connecting the moon with Venus in
the night sky. Similar bas reliefs of plant life and spirals as well as a massive
trilithon with a "V" shaped stone are found in other chambers of the temple.
Likewise, the doorways of stone with circles or small openings are reminiscent of the entrance to the womb-caves of the Paleolithic.

Finally, the design of the Ḥaġar Qim temple-complex itself is reminiscent
of the sexual metaphors associated with the goddess of regeneration as Venus. Like the Apse with the immense "V" sign in Lascaux cave and the
stalactite of the goddess and the aroused bull in the Chauvet Cave, the temple-complex of Ḥaġar Qim has a massive, "V" shaped stone in front of a
trilithon altar representing the vulva of the goddess. Moreover, the temple
itself is designed for privacy with its six separate chambers, three on the west
for the Evening Star and three on the east for the Morning Star. The three
western chambers each with a separate entrance oriented to three of the
positions of Venus and the moon when Venus is the Evening Star, intimates
the necessity of privacy for fertility and/or sexual rites. The three chambers
on the east are also chambers of a private nature. The stones used as doorways are small and separate the eastern chambers of the temple. Moreover,
the images of reclining goddesses, four of which were found at Ḥaġar Qim,

only serve to add to the overall image of fertility, sex for relaxation, spiritual renewal and the beauty and magnificence of the goddess in her form of the Regeneratrix of life. Like most of the temples on Malta, Ħaġar Qim represents the earliest form of worship of the goddess where man channeled the energies found in the sacred caves of the Paleolithic era into man-made replicas that served an increasing population who began to formalize and structuralize long-held beliefs in the sacred nature of the feminine.

Another noteworthy Neolithic passage temple possibly designed for the emulation and worship of the Regeneratrix in association with the planetary cycles of Venus, is Pierres Plates which is on the eastern side of the Locmariaquer Peninsula in Brittany, France. At Pierres Plates, the massive stones used for the roof of the chamber are on ground level. The passage temple of Pierres Plates is an underground chamber with a 120° angle or tilt like the tilt in the Apse at Lascaux. It is an allée coudée type of dolmen whose corridor changes direction and slightly enlarges to form a chamber, sometimes separated from the corridor by an upright stone marked with the symbols of the goddess. Like many of the temples of Venus, such as the temple of Ħaġar Qim at Malta and the temple of Aphrodite at Acrocorinth, Pierres Plates faces the sea giving it a spectacular view of Venus being born by rising from the watery realms. This temple, like Ħaġar Qim and the other Malta temples to the goddess, is aligned on a southeast-northwest direction with the entrance facing southeast. The temples of Malta are designed this way because they are shaped like the body of the goddess where her womb opens to the southeast and her head is represented in the northwest. At Ħaġar Qim the entrance faces the southeast and the rear faces the northwest with a broad view of the sea to site Venus as the Evening Star from the most reclusive temple chambers. The shelf at Les Eyzies-de-Tayac is also aligned on this axis and because it is a shelf, with a double view like Ħaġar Qim, Venus is seen rising on both horizons.

When the temple of the goddess, like Pierres Plates or Hal Saflieni, is an angled-passage, a subterranean chamber or a cave, the northwest siting is not visible from inside the structure unless there is a Shaft like the one in Lascaux cave. In these cases, the end chamber serves as an altar of the goddess; in Hal Saflieni, the end chamber has an altar that is a massive trilithon structure with a "V" stone in front of it, and in Chauvet cave, the end chamber contains a stalactite decorated with the vulva of the goddess. The apse or turning point to the altar is used as an oracle chamber or a private chamber in these passage temples. In Pierres Plates, the angle contains an apse as the turning point to a back chamber decorated with feminine signs of the goddess that, as one tourist states, "get more exotic the further you get into the monument." These chambers, which are aligned to the cycles of Venus, note both the planet's rising from the sea and the goddess' birth from the sea as well as its appearance in the evening sky as a symbol of beauty and fertility.

In the northwest chambers, Venus reaches her most spectacular beauty in the night sky and as the goddess, she is mounting The Columns of Life or The World Tree in her epiphany through the realms of her journey. In the east, the birth of the goddess is celebrated and in the west, the re-birth of the goddess is celebrated in the passage temples.

Besides the passage temples' alignments to Venus in the southeast as the Morning Star and Venus in the northwest as the Evening Star, the passage temples are aligned on the earth. Pierres Plates on the Locmariaquer peninsula, as well as Le Grand Menhir Briseé, the Kercado passage mound, Gavrinis and Carnac are all on the latitude of 47° north. The latitude of 47° north at Carnac is where the sun, at both the Winter and Summer Solstices, forms a perfect Pythagorean triangle relative to the east-west equinoxial axis of the site. Likewise, Le Grand Menhir Briseé, once the largest known standing stone in Europe, is also in the center of a circle of sacred temples. Le Petit Mont, a dolmen transformed by the Romans into a temple for Venus, is in the circle of temples that radiate from Le Grand Menhir. It is an eight level temple with several antechambers and stone carvings similar to those in Pierres Plates. These temples all share a common symbolism in the markings on the temples which are associated with the goddess as Regeneratrix. More than likely, these temples to the goddess are aligned on the earth with the necessary sitings to the solstices, equinoxes, the moon and Venus to complete a calendar which uses celestial cycles and a symbolic language as a guide to the infinite. They are aligned to the sky, the earth, and the subterranean world simultaneously, and they are united by a single symbolic language.

In Pierres Plates, the symbolism of the goddess as Regeneratrix forms a set of images and symbols that is abundant in the early Neolithic temples of France. In Pierres Plates, multiple pillars are decorated with large amounts of cupmarks and striated lines signifying the water aspect of the goddess of regeneration and the waters of life. Like the cupmarks and markings of wavy lines emanating from the center of claviforms on other Neolithic temples, water is a symbol of the renewable energy of the goddess as a force of new life, hence her birth and re-birth from the sea. On the stones at Gavrinis temple which depict an extensive use of wavy and concentric arc motifs radiating from a central "vulvar" opening, Gimbutas remarks that, "The piled-up signs seem to say that the creativity of the Goddess is inexhaustible and comes from the cosmic deep, which is implied by a variety of adjacent aquatic configurations" (*The Language of the Goddess* 225). This cosmology of water and birth as well as rebirth is an essential element in the Venus temples, like Pierre Plates, that are in proximity to the sea and the rising of the planet from the waters of creation both in the east and in the west.

Moreover, at Pierres Plates, there are thirteen stones with what appear as *figurations humaines* later identified as the "Pierres-plates style" of feminine figures which typifies almost all of the Morbihan entrance temples, according to a study by A. Maudet de Penhouet in 1814 (Whitaker 1). In this highly significant art form, the cupmarks or open circles and the wavy lines of emerging energy fuse to form a symbol of the figure of the goddess. Instead of a basin which is seen at the top of the four-sided pillar at Ħaġar Qim, the water element of the goddess becomes a wide-open "U" shape at the top of the goddess figure. The striations of water are replaced by a single line intersecting the "U" and running down the center of the stone to create the body of the goddess. Although the figure is unmistakably human in form, she is headless and abstract. The "U" formation and the extended bodyline of the goddess forms a configuration seen in the upraised arms of adoration of the Minoan-Mycenaean and Greek goddess and priestess figures and in the letter "ψ" (psi) in the Greek alphabet. It is also reminiscent of the figure of the goddess as Venus being raised from the sea in her foam-born birth on several depictions including those in Minoan art and the Ludovisi Relief where Aphrodite is raised from the sea by two nymphs. On the stones at Pierres Plates, the ψ is at the center of the stone dividing the stone into two spheres.

On one of the most decorated of the goddess figure stones at Pierres Plates, each side of the sphere contains four symbols; two symbols at the top and bottom of each column of four are concentric circles, and the two in the middle are half-circles. These might well be symbols for the phases of Venus in its full and crescent phases in its eight year cycle as the Morning and Evening Star. Moreover, at the top of this stone is a bowl shaped "U" that separates the two spheres of Venus with several lines indicating the crossover from east to west that Venus travels in its cycle. Not far from this stone is another indication that the passage temple of Pierres Plates is a Venus temple: it is a stone with a clearly marked Tree of Life with its branches reaching to the night sky. Finally, what emerges from the Paleolithic and early Neolithic depictions of the goddess of regeneration associated with the cycles of Venus, especially in this part of France, is a single recurrent symbol that fuses the goddess' attributes of renewing life from the uterine moisture of the feminine in the womb as basin and container of life and continuing cycles of feminine energy merged with a belief in the sacred. All this is contained in what the Greeks entered into their alphabet as ψ, what the Minoans represent in the adoration figures, and what the French eventually call Mari, perhaps the "maria" of "Locmariaquer."

The emergence of the goddess as Regeneratrix is depicted on the stones in the British Isles in an abstract method that is more accurate in detail than the abstract representations in France in its exactness of the cycles of Venus and in astronomy notations, overall. In Britain, the abstract symbols for the Venus cycle are intricate patterns that replicate the motions of the planet more than the symbols in France that merge the recurrent images associated with the Regeneratrix such as water, waterbirds, the tree of life, the "V" shapes and, finally, a human figure. However, in the emergence of the primordial archetype, the transformative element whether embodied in human form which eventually leads to language or embodied in a highly developed language of astronomy, dominates and identifies the archetype in both places. The transformative element of the primordial archetype in both cultures, is merged with one essential transformative characteristic of the archetype: the lunar cycle. Most prevalent is the merging of the archetype with the transformative symbols of the moon. For example, the Table des Marchand stone, discovered in the end chamber of the tumulus at Locmariaquer, is in the shape of a closed, pointed "U" with markings of the monthly, yearly, and precessional cycles of the moon. Likewise, at Knowth, one of three mounds of the Brú na Bóinne, a stone, called "The Calendar Stone" displays both The Nineteen Year Cycle of the moon as well as the eight-year cycle of Venus.

In Ireland, this is particularly true of many of the stones such as those discovered at the Brú na Bóinne which are unmistakably lunar-solar temples. The transformative character of the primordial archetype is therefore recognizable as both a spiritual and clearly observable essential in the understanding of the sacred feminine in conjunction with the moon. Moreover, as is the case with most accurate calendars, both ancient and modern, the lunar cycles and the cycles of Venus are also measured in conjunction with the cycles of the sun. These three celestial bodies, the moon, Venus and the sun, are essential to the forming of the transformative character of the archetype of the feminine because once again, they represent the powerful triad or trinity or what the Celts used as a central symbol in their religion: the triskele. In *The Stones of Time*, Martin Brennan remarks on the widespread distribution of this theme with the underlying assumption for its origins. Brennan states: "The separation of a formless unity into two reciprocal principles which generate a third and form the multiplicity of creation is a universal and archaic cosmological idea. In sky imagery the sun and the moon represent two opposite principles and the stars represent multiplicity. Together they make up time and space and the entire universe" (195). If the sun is represented as a masculine power to complement the transformative character of the sacred archetype of the feminine, then both the astronomy and the religious sentiment present a coherent ideology.

Brennan declares that the markings on the stones of the Brú na Bóinne and other Neolithic monuments form geometric patterns that essentially unite these realms. Brennan states: "The megalithic artist apparently viewed the multiplicity of the stars and the multiplicity of the 'world' below as originating from the same source, and both are seen to conform to basic geometrical structures" (195). Because there are only three celestial objects capable of projecting beams of light and casting shadows, Venus, the sun and the moon, Brennan believes that Venus, with its brilliant light, "became an object of awe in early astronomies" and its "path in the sky bears a close resemblance to the elongated arcs" on the stones (170). The role of Venus both as a goddess of regeneration and a planet, acts as a guide to uniting the feminine and masculine forces as she transcends from the realm of the earth through the waters to the night sky traversing the universe. No wonder the Greeks refer to her as Aphrodite Ourania, a goddess of astronomy born from the cosmic force of Uranus who marries Mother Earth to create the universe. Both the pattern of Venus in the night sky and her brilliant light form a constant, yet traveling celestial body that completes the cycles by uniting the feminine and the masculine in a coherent triad of the cosmos. In terms of astronomy, Venus is the force that is able to measure the accurateness of the lunar-solar cycle by appearing every eight years in exact conjunction with the moon and the sun, and in theological or spiritual terms, Venus acts as the impetus to uniting the feminine and the masculine as a cosmic force. Venus may be seen as a kelson of creation that holds the universal triad together.

The passage temples of the Brú na Bóinne represent the triskele which uses the unifying forces of Venus and its transformative properties in much the same way as other Neolithic temples. The shape and recesses of the temples, the measuring of celestial light in the passage, the associations with water and waterbirds, and the intricate patterns in the language of the stones all reflect characteristics of the goddess as Regeneratrix. Like the trilithon altars at the temples of Ḥaġar Qim and Hal-Saflieni in Malta, the passage temple of Knowth East at the Brú na Bóinne has a triple-alcove chamber. Likewise, the Newgrange passage temple of the Brú na Bóinne has a cruciform chamber with three recesses to represent the triad of Venus, the moon and the sun. Brennan cites three roofboxes in prominent Neolithic temples that mark the cycle of Venus on the entrance of each passage temple. The first is a sillstone at Gavrinis marking Venus in eight quadrangles, the second is the lintel at the entrance of Fourknocks, Ireland, using watermarks and quadrangles for the eight years of the Venus cycle, and the third is the roofbox at Newgrange (168). Brennan believes that "inherent in the structure of the quadrangle are the concepts of the center and the unification or reconciliation of opposites" because it is basically comprised of two triangles (183). Like the goddess herself as transformer of the feminine and masculine, her symbol is one of uniting opposites and is placed as such at the entrance to the temples.

According to Anthony Murphy in *Islands of the Setting Sun* (Dublin: Liffey, 2008), the bright light of Venus shines into the chamber of Newgrange once during its eight year cycle. That one moment of reconciliation of the cosmos, according to Murphy, is year four in the eight year cycle of Venus when the planet is the Morning Star at the Winter Solstice at the declination of Venus when it is 18° west of the sun (166). Here, the goddess lingers for several days before and after the Solstice in a different position in the roofbox at twilight where an observer would be able to see Venus through the roofbox if sitting on the floor of the chamber. This moment of reconciliation might have been marked by the ancients as a time of birth, re-birth and regeneration typical of the functions of the goddess as Regeneratrix. According to Murphy, the word "Brú" may relate to "the notion of a 'rebirth ritual' coincidental with the solar and lunar functions of Newgrange" (172). Murphy goes on to explain that the Brú na Bóinne complex might be considered The Womb of the Moon where the temples take the form of the body of the goddess (172-73). The light of Venus shining into the roofbox would play the role as a powerful Regeneratrix for The Womb of the Moon when the light of the Sun as a masculine energy force at the Winter Solstice enters the womb temple. The two phallus-shaped stones in oval settings filled with white quartz that were discovered at the entrances of Newgrange and Knowth contribute to this theme. Here, as in the caves of the Paleolithic and the temples of the Neolithic, the entrance is the vulva of the goddess and the passage chambers act as temples where re-birth and fertility rites most likely took place.

At the end of the cruciform chamber at Newgrange is a basin used for ceremonial rituals most likely in association with the water aspect of the Regeneratrix. According to Murphy, Joseph Campbell cites a local legend about a "morning star" that cast a beam of light into the basin at Newgrange. Murphy believes that this local legend might well be an enduring myth perhaps as old as five millennia (165). Noticeable too at Newgrange on several stones on the outside of the passage are cupmarks and other water symbols of regeneration and rebirth from the watery realms of the goddess. Most notably, is Murphy's research on the nearby site of Baltray where Venus is seen rising from the sea. Although the major connection with water as a sacred source of birth and re-birth at Newgrange is the Boyne River that gracefully flows around the Brú and whose later namesake is Boann, the river goddess of the Tuatha Dé Danaan in Celtic mythology, the passage temple is also associated with another sacred water source: the sea at Baltray. Murphy remarks that Venus is a female deity represented in Celtic mythology as a calf whose mother cow is the moon, and both rise from the sea at Baltray where they have been transformed into stones. Venus rises from the sea in the southeast at the time of the Winter Solstice not far from the Rockabill Islands at Baltray in her new sacred form (11-15). Like the sitings of Venus rising from the sea at Malta and in France, Venus rising from the sea at Baltray is marked at the Winter Solstice, as it is at Newgrange, as a sacred re-birth.

Although one animal associated with the waters of the goddess at Baltray is a white cow and her calf, and the same association with bovine figures is seen in connection with the goddess of the Boyne River, Boann, these later Celtic myths also include an association of the water sources with waterbirds which may be traced to the regenerative rather than the elementary character of the archetype. The bovine figure, which enters into the Bronze and Iron Age mythologies of the Celts and the Greeks, in the form of Boann and Hera, respectively, may be said to represent the elementary character of the feminine archetype which clearly emerges in later mythologies in bovine form with the domestication of animals. The transformative animal form associated with the sacred feminine is the bird, a creature seen as early as the Paleolithic as a force that transcends ordinary human consciousness by transcending from the watery realms to the sky. This form most closely associated with the flight of Venus from its rising as the Morning Star in the eastern twilight, is represented at Newgrange by the whooper swans (*Cygnus Cygnus*) who winter at the mound in significant numbers, the marking of the constellation of Cygnus as the shape of the passage of the mound, and the later associations of the Celtic goddess Caer who is transformed into a swan and dwells at Newgrange. Moreover, the tales of Dechtine, who births the famous Celtic hero, Cúchulainn in the mound, and the Children of Lir who transform into swans at the mound adhere to the waterbird significance at Newgrange (Murphy 142-48).

Nearby Newgrange, there are two interesting sites to include in the sacred sources for water and waterbirds in association with the Regeneratrix. Both are mentioned by Murphy as the man-made circular ponds at Newgrange and the "Newgrange-Fourknocks harmonization" (148-152). Along the Boyne River not far from Newgrange, there are four circular ponds, reminiscent of the man-made wells at Ħaġar Qim, where swans winter and feed in the shallow waters. Murphy notes that the swans were "held in great reverence" as was the constellation of the giant swan as Cygnus. In addition to these water sources and waterbirds, the nearby passage temple of Fourknocks reflects the same motifs in its structure and artwork. The egg-shaped chamber of Fourknocks might be seen as a cruciform design with three recesses, much like those recesses in other regeneration chambers. Fourknocks also has zigzag water patterns that mark the passage temple as one of regeneration as does its central lintel, the west recess lintel, and the entrance lintel with eight quadrangles for the Venus cycle (Brennan 185). According to Murphy: "Of significance is the fact that both Newgrange and Fourknocks have as an integral element of their design, the shape of an egg, the symbol of fertilization, of reproduction, of new birth. We are reminded of the fact that the primordial cosmic egg has its roots in much more ancient times, and is connected with symbolism which includes water, a bird, a woman and an egg. The woman in this case is Caer" (152).

The passage mound of Knowth at the Brú na Bóinne may also be associated with birth and re-birth in the cosmology of the sacred waters of the feminine because it contains a basin in its cross-shaped eastern chamber. Knowth and Dowth, which combine with Newgrange to make the sacred Brú, according to Murphy, "utilized differing methods to evaluate, enumerate, and predict the same thing—the standstills, or 'lunistices' of the moon" (195). Their connections to the cycles of Venus may be seen on the stones and, in particular, in the markings on the stone basin at Knowth. The stone basin found in the East Passage of Knowth is decorated on the outside with a pair of concentric arcs with a circle in the center of each. It seems as if there are eight concentric circles with a spiral dissecting the fourth circle on the left. The arc may well be a symbol of the rising of Venus, and the fourth year marked by the spiral may well be the rising of Venus in the Winter Solstice twilight of the fourth year in the eight year Venus cycle. On the bottom of the stone basin is a circle with arcs that resemble halos around the center circle with six radiating lines on each side of the space beneath the circle. Whether they are configurations of the Venus cycle is speculative; however, the markings in the basin are clearly associated with water and the radiating energy produced by water as a sacred life form. The fact that the basin is in the East Chamber of Knowth is also indicative of the birthing aspect of the waters of the feminine.

Finally, and probably most dramatic of all the evidence of the Venus cycle and the Regeneratrix in the Brú na Bóinne, are the markings of cosmic cycles on the kerbstones of the mounds. These markings depict the cycles of Venus, the moon, and the sun in an intricate art form that has not been superseded in detail and accuracy in any other culture proving the ancient astronomers of Ireland masters at their art. At Newgrange, the Entrance Stone, the stones in the passage chamber, and Kerbstone 52, which is half way around the outside of the mound, are decorated with the symbols of Venus, the sun and the moon in varied patterns that seem to express the merging cohesiveness of the cosmos in all its complexity. According to Murphy, the astronomical alignment of these stones is of central consideration to the monument because they mark the light of Venus, the moon, and the sun at the Winter Solstice as they enter the mound, reach the back chamber and pass over Kerbstone 52 (168). The Entrance Stone depicts this cosmological triad with triple-spirals and quadrangles in complementary patterns on either side of a central line of a giant egg-shaped stone that guides the light into the mound; the roofbox with the Venus lintel that guides the light of Venus is positioned directly above the dividing line of the Entrance Stone. As the light passes through the chamber, stones decorated with triple-spirals are illuminated. Kerbstone 52, on the outside of the mound, is also bisected with patterns of the triple-spiral and quadrangles as well as cup-marks which suggest the watery element of the goddess as Venus.

On the passage temple of Knowth, Kerbstone 52 has been called the Calendar Stone by Martin Brennan. This stone which combines a spiral, wavy lines and a series of circles and crescent shapes depicts the Nineteen Year Cycle of the moon. According to Murphy: "The numbers and arrangements at Stone Age sites were chosen so that there were several ways of counting them. Thus, the Calendar Stone can be counted as a 62-month metonic interval, or a 99-month eight-year Venus-moon interval" (197). Likewise, at the mound of Dowth, Kerbstone 51 on the eastern side of the mound has sundial shapes containing stars with radials that count 99 months which brings the eight year cycle of Venus together with the cycles of the sun and the moon. Brennan states that on Kerbstone 52 the eight circles and ovals very likely represent the eight years which are reduced to terms of months on Kerbstone 51 (166). On Kerbstone 52, eight circles and ovals of varying shapes are decorated with small black circles. The pattern of the circles and ovals resembles the journey of Venus where it may be sited in the east on three occasions in relation to the moon, in the west on three occasions and below the horizon in two positions. Like many of the patterns of Venus on ancient monuments of the Brú na Bóinne, Venus is seen as a vital part of the cosmic unity of Venus, the moon, and the sun at Dowth.

A relatively small passage temple in the British Isles that indicates a purpose similar to the temple of Ħaġar Qim in that the pattern of the temple displays more qualities of a Venus temple rather than the grand passage temples of the Brú na Bóinne which represent a cosmic calendar of magnificent proportions, is the passage temple of Bryn Celli Ddu on the Isle of Anglesey in northern Wales. Bryn Celli Ddu, like Ħaġar Qim and Pierres Plates on the Locmariaquer Peninsula in France, faces the sea. Bryn Celli Ddu is approximately a mile from the channel that separates the Isle of Anglesey from the mainland of Wales with a dramatic view of the water from the typically steep stone cliffs off the coast of northern Wales. The opening to the chamber of the mound faces northeast; however, a slot or opening, reminiscent of the oracle chambers in the passage temples of Malta and the Shaft at Lascaux, faces the southeast with a view of Venus from the chamber. Two pillars, one on the outside of the passage temple and one free-standing pillar in the chamber, may be seen as symbols of The Columns of Life or The World Tree. The outside pillar is decorated with waving lines of watery energy and the inside pillar, according to Christopher Knight and Robert Lomas, marks the light of Venus: "Around the winter solstice, the slot and the pillar of the chamber accurately measure the angular distance of Venus from the sun, using the difference between daggers of light cast by the sun and Venus onto the pillar. The positioning of the slot has been carefully designed to make this possible." According to Knight and Lomas, in Welsh mythology, this passage temple is associated with the mythic astronomer Gwydion ap Don, the son of Don, the mother goddess of Wales (234).

Whether grandly overpowering and magnificent, as seen in the passage temples of the Brú na Bóinne and Ħaġar Qim, or startlingly remote and private, as seen in the passage temples of Bryn Celli Ddu and Pierres Plates, the connection to the starry dynamo of energy worshipped as the regenerative forces of the sacred feminine is evident in the temples of the Paleolithic and

Neolithic eras in the history of human consciousness. The development of an emerging archetype given the characteristics of birth and re-birth, avenues for the immortality of the spirit to transcend this reality to other worlds, is worshipped in distinct forms that guide our imagination. The symbols of birth and re-birth with Venus rising from the watery realms as the Eastern Star, the waterbird as a messenger and guide of the immortal soul, the journey from east to west and then to the world of darkness and sleep, and the mounting energy of the planet as it traverses the night sky seen as a sexual metaphor contribute to the overall complex of images of the transformative character of the goddess. The Columns of Life, The World Tree and The Axis of the World that the goddess mounts with energy and glory, all intensify the experience of the fertile and nubile feminine in the universe. Most of all, the abundance of female symbols and the emerging language of the goddess that crystallizes into the language and worship of the organized religions of the Minoan-Mycenaean civilizations that follow becomes an overwhelming symbol of the feminine in human history. Known as Regeneratrix in her transformative character, she is seen in the night sky as Venus.

WORKS CITED

Brennan, Martin. *The Stones of Time: Calendars, Sundials, and Stone Chambers of Ancient Ireland*. Rochester, Vermont: Inner Traditions, 1994.

Campbell, Joseph. *Mythologies of the Primitive Hunters and Gatherers: The Way of the Animal Powers*. Vol. I. New York: Harper and Row, 1988.

Eliade, Mircea. *Myths, Rites, Symbols*. New York: Harper and Row, 1976. Gimbutas, Marija. *The Language of the Goddess: Unearthing the Hidden Symbols of Western Civilization*. New York: Thames and Hudson, 1989.

Hesiod. *Theogony*, Trans. H.G. Evelyn-White. *Theoi Greek Mythology: Exploring Greek Mythology in Classical Literature and Art*. New Zealand: The Theoi Project, Ed. Aaron Atsma, 2008.

Leroi-Gourhan, André. *Treasures of Prehistoric Art*. New York: Harry N.Abrams, n.d.

Knight, Christopher and Robert Lomas. *Uriel's Machine; Uncovering the Secrets of Stonehenge, Noah's Flood, and the Dawn of Civilization*. Glouster, Mass.: Fair Winds Press, 1999.

Murphy, Anthony and Richard Moore. *Island of the Setting Sun: In Search Of Ireland's Ancient Astronomers*. Dublin: The Liffey Press, 2008.

Neumann, Erich. *The Great Mother: An Analysis of the Archetype*. Princeton: Princeton Univ. Press, 1974.

Thurman, Judith. "Letter from Southern France: First Impressions" *The New Yorker*. 23 June 2008: 58-67.

Whitaker, Alex. "Les Pierres Plates" *Ancient Wisdom*. 2010. 13 February 2011. http://ancient-wisdom.co.uk.

Chapter Two

The Epiphany of the Goddess

A Study of Venus in the Bronze and Iron Age

Helen Benigni

The developing iconography of the feminine in the Minoan-Mycenaean culture follows the ideology of the Paleolithic era and Neolithic era in the sense that it exhibits both the elementary and the transformative characters of the archetype. In the Minoan-Mycenaean culture, the statues and frescoes at Knossos and Mycenae depict two distinct figures associated with the feminine and, in particular, with the religious icons of goddesses displaying the elementary and transformative characters of the archetype. The elementary character of the archetype is seen in figures that assume a posture of what traditionally has been defined as an earth goddess, here distinguished by her rounded figure and arms that cup her breasts as well as her association with plants, and in particular poppies, the earth and the underworld. Archeologists at the Mycenae site museum term these figures "phi" goddesses because of the eventual evolution of the figure into the letter of the Greek alphabet "Φ." In Minoan Linear B, the ancient script of the Minoan-Mycenaean culture, this goddess is referred to as The Mother of the Goddesses and Gods or the pre-Greek Demeter and is often depicted with a dual nature of mother-daughter as Demeter and Poseidonia or Ariadne, two goddesses who are possible models of a pre-Greek Persephone (Gimbutas 145-46). In Minoan-Mycenaean culture, her animal totem is often the sow, the boar, or the magical griffin, a combination of the earth powers of the bird and the lion with the human imaginative powers, and she is associated with wheat, pomegranates and poppies.

The transformative character of the archetype is likewise depicted by a figure in the Minoan-Mycenaean script and Greek alphabet as "psi" represented by "Ψ." The transformative character or "psi" character of the archetype with arms raised in a posture of adoration of the heavens is associated with the changing cycles of the moon and Venus, the two celestial bodies honored in a practical sense to determine calendar time and honored in a religious sense as the powers of the goddess to impart spiritual enlightenment through celestial awareness. Although all figures of the goddess, both elementary and transformative in their character possess a certain degree of the sacred deemed sacrosanct by their nature to represent eternity and eternal love, the transformative character serves as sacred, particularly, in its representation of time and the cycles of the cosmos. As seen in the Paleolithic era and the Neolithic era, the transformative character of the feminine archetype represents both the cycles of the moon and Venus in conjunction with the sun. In mythology, this translates as the two interchangeable or seemingly monotheistic goddesses known in Minoan-Mycenaean and Greek culture as Hera, the goddess of the moon and the bull, and Aphrodite, the goddess of Venus. The goddess of the yearly cycles of the moon and the cycles of precession is often depicted in the same costume and on the same signet rings and frescoes as the goddess of Venus as their functions would require an association of almost interchangeable characters. The moon goddess, however, is almost always associated with the bull, the bull's Horns of Consecration and The Sacred Axe while the goddess of Venus is determined by her association with water, lustral basins, birds, The Tree of Life, and most importantly, the cycles of the planet Venus.

In Minoan-Mycenaean iconography, as in the Paleolithic and Neolithic iconography, the transformative goddesses are often seen in a triad which establishes the sacred powers of the goddess. On one signet ring from the Mycenaean Cult Centre, Aphrodite is depicted with Hera and The Mother of the Goddesses and Gods and each is identified by the symbols of the planet Venus, the cycles of the moon, and poppies, respectively (Evans 10). On this particular ring, Aphrodite and Hera stand together surrounded by six bull's heads, The Sacred Axe, the figure eight shield of a Minoan god, who is most likely a pre-Greek Adonis named as *kouros* and Hyakinthos in the Linear B Pylos tablets, and their celestial symbols hovering overhead. They face the Demeter/Persephone prototype who is seated beneath a fruit-laden tree holding three poppies. Likewise, on "The Ring of Minos," the ring of the High Priestess of Knossos, three forms of the goddess are depicted as Hera holding The Horns of Consecration with Aphrodite, again on the left, facing a Persephone-type goddess who journeys through the Underworld on a seahorse boat. On this ring, a nude Aphrodite climbs The Column of Life, in this particular case most likely represented as Aphrodite's plant atop a column, in an ecstatic state; eight bulbs each in a progressively more blossoming state represent the eight years of the cycle of the planet Venus, and the goddess' journey is represented by her climbing, descent and journey to the Underworld.

In addition to the identifiable forms of the elementary and transformative character of the archetype of the feminine as well as the identifiable forms of the triad of the goddess in a developing pantheon of goddesses, the Minoan-Mycenaean culture seemingly portrays a bi-level to the pantheon's organizational structure. In addition to what might be called mother goddesses or more accurately, predominant goddesses, such as Aphrodite and Hera, are minor figures in the pantheon that are indicated as such by their significantly smaller size. These figures, also represented in Linear B by their actual names, are frequently depicted in Minoan-Mycenaean art beside the predominant goddess figures. The figures are most likely the daughters or less predominant figures of the pantheon. As both predominant and minor figures are also priestesses who are earthly representatives of the goddess, these figures indicate the relationships and possible rituals associated with the worship of the goddess. It might also be noted that male figures in Minoan-Mycenaean art are almost always smaller than the predominant goddess figures. In the case of Aphrodite, the male figures are most likely the representative of Hyakinthos, Adonis or an Adonis-like *kouros* as in the case of the male figures in the fresco at the Mycenae Cult Centre. In the case of the smaller female figures, they are most likely Britomartis, Aphaia, Diktynna, or Eileithyia who are all goddesses mentioned in Linear B Tablets at Knossos and Pylos.

In Minoan-Mycenaean art, the goddess of Venus is most easily identified by several features, the basis of which finds its roots in the Paleolithic and Neolithic eras. The goddess' association with water in the Paleolithic era and the Neolithic era is continued in the Minoan-Mycenaean culture with the introduction of the use of lustral basins in the southeast corner, in the northwest corner and in the Throne Room of the temple-complex at Knossos in Crete. At Mycenae, the lustral basin is found in the oldest section of the complex that has been titled the Cult Centre which is located to the east of Grave Circle A at the base of the complex. These site locations are, of course, oriented to Venus rising in the southeast and in the northwest. The goddess' association with waterbirds is also retained throughout the Minoan-Mycenaean iconography using the dove, the goose and the duck as her totem birds. The Tree of Life and the goddess' association with The Columns of Life are easily identifiable at Knossos with its Pillar Crypts which are located near the lustral basins and at Mycenae in the Cult Centre. In Minoan-Mycenaean art, the sacred pillar or tree is often represented as the fennel plant atop a column, as the fig tree, as the myrrh tree and as the myrtle tree. These sacrosanct replacements of the Paleolithic stalactites evolved from and were used simultaneously with the sacred cave sites such as the cave for Eileithyia at Amnissos on Crete.

Most important is the continued use of the goddess' epiphany or sacred awareness of the rites of fertility and the inspiration that the goddess provides as the source of regeneration both spiritually and physically. For this, the Minoan-Mycenaean culture depicts the goddess of Venus as the golden one; the Greeks continue this reference in several sources one of which is Homer in his "Hymn to Aphrodite" where she rises from the sea and is clothed by the Hours with a gold crown, gold earrings, and gold necklaces (*Hymn VI*). Her association with regeneration is seen in the gold ornaments and artwork which symbolize her brilliance, bright light, and sacredness. As nymph, bride and Regeneratrix, the goddess is part of the cosmic creation cycle of birth and re-birth. Her mobility and direction as well as her proximity to humanity are represented in her physical journey through the night sky as the planet Venus. Here, the Minoan-Mycenaean culture uses the cone as the symbol for the planet and the goddess' journey through the night sky. The cone adorns her headdress and is used in many of her cult associations as an indicator of her rising in the southeast as the Morning Star, her ascending in the north-west as the Evening Star, and her descent below the horizon in her search for her Adonis in the Underworld in her final phase. These patterns in the sky identify the goddess as a life energy that reaches a peak, or point in a cone, to disperse sacredness to the *hieros gamos*.

The right side of the cone represents her eastern voyage, the left represents her western voyage, and the base of the cone represents her journey below the horizon. The point of the cone may be seen as her pinnacle of ascension or her epiphany. Here is where the goddess mounts the cone and reaches the ascent of her experience of cosmic and sacred love. On the gold ring of the High Priestess at Knossos, she is naked, reaching for a fruit from The Tree of Life in an ecstatic state. On her descent from this peak experience, the goddess' head has been replaced with antennae, a symbol of her transcendence. On a gold pin from Grave Circle A at Mycenae, eight magnificent floral designs blossom from her head and majestically end at her feet. And, on a gold signet ring from Isopata, Crete, three priestesses are involved in an ecstatic state of adoration with antennae replacing their heads as a symbol of their reaching an epiphany. The goddess' bird is associated with her epiphany and is often seen in the artwork as a representation of her soul reaching its height through flight; two nude goddess figures made of gold leaves were discovered in Grave Circle A at Mycenae adorned with birds taking flight which are attached to their heads and elbows. Beside the bird in this journey, the goddess is accompanied by a young male lover later identified in Greek mythology as Adonis. In Greek mythology, the goddess of Venus must chase her lover into the Underworld as part of her journey and share him with the goddess of the Underworld, Persephone, for one third of her cycle (Apollodorus 3.14.4).

By sharing Adonis with Persephone, the goddess of Venus is able to retrieve love from the Underworld and renew herself for another journey. Like the precessional cycles of Hera who must renew her energies with the sacrifice of the bull, a symbol of the renewal of lunar forces both within the year and in the nineteen-year cycle of the moon, Aphrodite must renew her powers of love and regeneration by retrieving Adonis from Persephone and the Underworld. This emerging myth seen in the frescoes and the artwork at Knossos and Mycenae is one of two myths that define and identify the goddess of Venus by associating the iconography with the later Greek myths. Seen first in the Minoan-Mycenaean culture as a myth explaining the cycle of the planet as a cone or journey where two goddesses evidently share a young male as consort is one representation of the goddess that is repeated throughout Greek mythology. The other image that evolves from the Minoan-Mycenaean culture is the birth of the goddess as the Eastern Star rising from the waters of the Underworld. This, too, is an essential myth of the Bronze and Iron Age Greeks. Here, Aphrodite is most clearly associated with the waters of rebirth and renewal in her southeastern lustral basin rooms of the Minoan-Mycenaean temple-complexes; she is accompanied by nymphs as she rises from the sea like the planet itself rising from the Mediterranean Sea.

The earliest evidence of the worship of the goddess of Venus is found in the Paleolithic caves and mountain sanctuaries of Crete. Caves that contained chambers, passages, stalagmites and stalactites, and wells of pure water are those most closely associated with the goddess. Marija Gimbutas remarks that "the caves themselves made more dramatic and powerful religious sanctuaries than crypts, with their often remote settings, fantastically shaped stalagmites and stalactites, and shadowy interiors illumed only by torches and lamps" (138). Some temple-complexes, like the Knossos temple-complex, maintained their own sanctuaries one of which is the cave of Eileithyia at Amnissos. This sanctuary, mentioned by Homer in *The Odyssey*, is located in the harbor town of Amnissos with a clear view of the sea. Inside the entrance there is a cylindrical stalagmite enclosed by a quadrangle setting of smaller stones with one large quadrangle stone in front of the stalagmite (Nilsson 58). Like the altars at Hal-Saflieni and Haġar Qim, this altar has a similar shape and function and appears to be an altar to the goddess of birth and re-birth in this case named Eileithyia. Cult items such as a bathing tub found at the sanctuary cave of Mt. Ida and an altar stone with a ladle cup in its center as well as figures of birthing women found at Mt. Juktas, a peak sanctuary adjacent to the Knossos temple complex, attests to the fact that the goddess as Regeneratrix had a cult following. Mt. Jutkas, with its triangular terraces, provided the astronomer priestesses with an optimal place to view the movements of the celestial bodies such as Venus as did the cave at Amnissos with its view of the sea.

The Knossos temple-complex reveals evidence of the worship of the goddess as Regeneratrix with its lustral basins and chambers in the Southeast section of the temple; in the Throne Room and The Shrine of the Dove

Goddess at the center of the complex; and, in the Northwest Lustral Basin
Chamber and outdoor amphitheater. However, the temple-complex itself was
built as a temple to all the deities of the Minoan-Mycenaean pantheon.
Chambers and smaller temples within the complex are designed for the ritual
re-enactment of the myths of the religion and the design of the temple is a
conglomerate or labyrinth of several smaller temples devoted to spiritual and
secular life. The three circular depositories in the front of the temple were
most likely associated with the worship of the elementary archetype of the
earth goddesses represented as Demeter and Persephone. The sacred olive
grove and dance floor were most likely used for the rituals and worship of the
goddess as Hera as was the priestess' villa with its northeast siting terrace for
lunar, solar, and stellar observations. The abundant amount of Horns of Consecration situated throughout the temple-complex as well as the pillar crypts
and the avenues and terraces devoted to the bull-grappling, bull-netting and
bull-jumping events are evidence for the worship of the goddess Hera and the
sacred bull sacrifice in honor of the yearly and precessional cycles of the
moon. And, of course, several temples are dedicated to the Regeneratrix
through the ritual worship of the Venus cycle and its related myths.

The Southeast section of the East Wing of the Knossos temple-complex
contains the most open and sunlit rooms in the complex because of its four
massive skylights, numerous windows and orientation to the southeast and
rising of Venus as the Eastern Star in the early mornings as well as its use of
larnakes or bathing tubs and a lustral bathing room. The birth and re-birth of
the goddess rising from the sea is depicted on the larnax discovered in the
temple chambers, one of which is the goddess rising from the waves as an
octopus. The fresco of the "Ladies in Blue" was also discovered in the east
wing with its doors facing due east much like the eastern façade of the Haġar
Qim temple to the planet Venus. A similar fresco discovered in the east wing
of the temple at Akroteri on the island of Santorini, called the "Fresco of
Crocus Gatherers," pictures a goddess seated high on a throne on a tripartite
platform with young girls bringing her baskets of flowers. Flowers and floral
designs amidst dolphins also decorate the frescoes of the Southeast section of
the Knossos temple-complex. A sacred chamber called the "Lustral Basin"
faces Southeast and was most likely used as a sacramental basin for anointing the body with oils and water in re-birth rituals as the basin did not drain
and was made of materials inappropriate for constant use of water. According to Martin P. Nilsson, this basin was situated in a part of the temple-
complex where a shrine was set up which included a statue of a goddess with
a dove on her head (78). Like the lustral basin at the Phaestus temple and the
cave sanctuaries, this basin was most likely used for sacred water rituals
because of its quadrangular basin and central column design (Nilsson 94).

In the Northwest section of the Knossos temple-complex is another Lustral Basin of similar design and composition. It is located in a building that is

on the processional entranceway and is directly opposite an outdoor amphitheater and an outdoor fountain and shallow pool. Most likely, this Lustral Basin was also used for purification rituals related to the goddess of regeneration because of the use of a column within it as a representation of The Tree of Life, the sacred vessels discovered in the chamber, and the location of the building for the northwest siting of Venus as the evening star. The proximity of the theater to the Lustral Basin and to the ceremonial entrance or processional road called the Royal Road suggests that the three structures were used together as an area for ritual initiation and purification for those first entering the temple-complex from the main road that leads to Herakleion, a city on the northern coast of Crete. Venus as the Evening Star is clearly visible from the seats in the theater, and the addition of the pool to the Lustral Basin provides public access to another ritual purification source much like the sacred wells outside the temple of Haġar Qim and the natural access to the sea at many of the Neolithic temples to the goddess as Regeneratrix. Quite possibly, the theater might have been used to present the myths of the goddesses, most notably the myths of the Minoan Aphrodite rising as the Evening Star and the sharing of the *kouros* or Hyakinthos between the Minoan Persephone and Aphrodite.

Although the Lustral Basins in both the Southeast and Northwest sections of the Knossos temple-complex identify the goddess of regeneration with water and the renewal of life, the Lustral Basin in the center of the complex in the room known as the Throne Room is the most impressive and convincing dedication to the goddesses worshipped by the Minoan-Mycenaean civilization. Here, the goddess in her state of epiphany is realized by the representations of griffins in frescoes, statues of the goddess in the adorant or "Ψ" position, a lustral basin with four large windows facing due east, and tripod stands which presumably held sacred vessels for anointing. Most importantly, the Throne Room's entrance was a Tripartite Shrine with five pillars perhaps signifying the five positions of Venus in the night sky. The Minoan Aphrodite is clearly seen as a vital figure in the Minoan pantheon with the representations of the sacred bucrania and Horns of Consecrations devoted to Hera as well as those icons devoted to the Mother of the Goddesses and Gods or the Demeter/Persephone figure of the pantheon. Not far from the Throne Room are the Pillar Crypts containing the sacred pillars as The Columns of Life and the sacred *xoana* which were most likely ritually bathed and displayed as they were at the ancient Greek festival of the Aphrodisia at Athens and the Daedala at Argos as well as many of the festivals at the smaller Greek temples of goddess worship.

The griffins which flank the seat or throne of the High Priestess and the doorway to the Interior Shrine, and the statues of the goddess are representations of the epiphany of the goddess as a combination of several of her characters in one form. The body of the griffin is the lion which represents the earth or elementary character of the archetype; the head of the griffin is the bird with its five spiral plumes which represents the celestial and trans-formative aspect of the goddess; and, the watery base of the griffin represents the underworld from which the goddess journeys in a continual birth and re-birth cycle. The many statues found on the altar of the Interior Shrine of the Throne Room, like the statues found in many of the temples of the Minoan-Mycenaean temple-complexes, are also a combination of icons representing the epiphany of the goddess. The statues identified with the goddess as Regeneratrix have three symbols on their crown. Their crowns often contain birds of the goddess representing her spiritual and celestial state of epiphany as the goddess of Venus, two horns of consecration representing her state of epiphany as the goddess of the moon, poppies representing her Persephone nature, and a cone in the center representing the entire journey of her epipha-ny. On these crowns, the base of the cone represents the 55-60 nights that Venus is under the horizon presumably in her journey to retrieve her *kouros* from Persephone and the Underworld; the right or east side of the cone represents the 260 mornings the goddess is born and re-born from the Under-world as the Eastern Star; and, the left or west side of the cone represents the 260 nights the goddess is the Evening Star.

Sir Arthur Evans makes accurate observations and assumptions concern-ing the cone of the Minoan Aphrodite in his text entitled *The Mycenaean Tree and Pillar Cult and Its Mediterranean Relations* (London: Macmillan, 1901) despite the fact that Evans does not connect the cone iconography with the movement of the planet Venus, even though he associates the cone with Aphrodite Ourania or her equivalent in several cultures. Evans remarks that the cone is combined with the base of the pillar or tree when used as a baetylic table of offering based on his observations of the offering table from the Dictean Cave, depictions of baetylic altars on Cretan coins, and the sacred cone at the grave of Aphrodite on Paphos (16-22). Evans uses the Cone of Astarte from Byblos, and what he describes as "a double cone in reversed positions" found on Babylonian cylinders as well as double cones used as bases for altars as a means of comparison and a demonstration of the influence of the Babylonian culture on the Minoan-Mycenaean culture (40; 53). The double cone may represent the completion of the four year cycle of Venus when it returns to the same lunar cycles in a four year and eight year period. Evans then identifies the goddess associated with the cones and makes note of the celestial imagery connected with the goddess' image: "The star and crescent, the rays which generally issue from the stone itself, point to her in her character of a luminary of the heavens, Aphrodite Urania" (74).

Evans also comments on two altars which he believes offer the most complex composition of the shrine to the Minoan-Mycenaean goddess, one which was found at the Acropolis of Mycenae in Grave Circle A and the other which was found at the Knossos temple-complex. The Triple Dove

Shrine from Mycenae is an altar with doves perched on either side of a three-column base toped with the Horns of Consecration. This shrine seems to be a miniature temple to what Evans sees as "the pillars of the house" which are raised upon a stonework base and set on the roof of the central shrine (93). The altar at Knossos is a shrine of three pillars with a dove atop each pillar discovered in what was believed to be a sanctuary or temple above the Throne Room at Knossos. The doves rest above two spheres as nests for the altar to the Minoan Aphrodite. Perhaps the spheres, like the two sides of the cone, represent the journey of the goddess in the eastern and western sky. In this depiction, the goddess as Venus is represented at the highest part of the Knossos temple-complex indicating her dominance as a celestial deity. According to Nilsson, this pillar shrine was believed to be in a room above that had collapsed but was once thought to be a sanctuary. In it was the dove shrine with seven columns of different sizes, "a model of a portable seat," fragments of miniature triton shells of painted terracotta, and a series of miniature vessels. Nilsson calls this miniature temple The Shrine of the Dove Goddess (88).

Marija Gimbutas has perhaps the most accurately imaginative vision of what the rites and meanings associated with the Throne Room and its adjoining chambers and sanctuaries might have meant. Beginning with the Throne itself, Gimbutas remarks: "The seat back of the throne has curving sides decorated with semicircles and a protrusion on top (it is actually roughly anthropomorphic)" (137). On its base is "an omphalos with the full moon: symbols of concentrated life force." Gimbutas sees these symbols as a continuing iconography inherited from earlier times. She describes the Throne Room as such:

> The "throne room" obviously served ceremonial, not secular purposes. We can easily envision the rituals that took place here. In the adjoining preparation room, the priestess was dressed in festive, symbolic attire. She appeared at the door, flanked with sphinxes, and then advanced to the throne, where she received offerings brought through the same door from the service station (a suite of rooms that included tables, benches, and low stone seats next to the preparation room). The priestess became the earthly representation of the goddess.
> A lustral basin stood across the room and down several steps. Its presence reveals a ritual that may have included descending to be cleansed with water and fragrant oils, and ascending again, renewed.
> (137-38)

Certainly, Gimbutas captures the essence of the goddess as Regeneratrix in the form of either the moon goddess and/or the goddess of Venus. The Throne Room serves as possibly the best example of the epiphany of the goddess for the rituals and myths associated with the Minoan-Mycenaean religion. However, it was not the only temple-complex of magnitude in the Minoan-Mycenaean cultural sphere.

The Cult Centre of Mycenae which is situated below the acropolis to the east of Grave Circle A, where the oldest remains of the Mycenae temple-complex have been unearthed, is vastly similar in structure and purpose to the temples of the goddess of regeneration at Knossos, Ħaġar Qim, Ħal-Saflieni, Newgrange, Pierres Plates, and Lascaux. Perhaps because it was relatively undisturbed when the acropolis was constructed above it and because of its position inside the Cyclopean walls of the Mycenaean fortress, or its proximity to the graves of the ancestors, it was left relatively intact as a fundamental shrine to the siting of Venus as the Morning and Evening Star. Apparently, little or no evidence of this sort remains on the northwest side of the acropolis of Mycenae due to its fortification of massive walled structures that protect its water source which was constructed as an invulnerable underground fountain. However, at the southeast section below the acropolis is the Cult Centre of Mycenae, a temple to Venus complete with two separate structures for siting Venus in the southeast and northwest.

In the Cult Centre of Mycenae, the two temples, one in the southeast section of the Centre and the other in the northwest section of the Centre, are located to the left of Grave Circle A, the burial grounds of the ancestors of Mycenae. The temple in the southeast section, named the Tsountas House after the archeologist who excavated it in 1893, is linked by a circuitous processional way to the acropolis above. At the bottom of the steep hill of the acropolis, the processional way makes a sharp left turn into the Cult Centre and leads directly to the Tsountas House or Southeast Temple. The Southeast Temple has a plastered square altar in its forecourt and a horseshoe shaped fixture in its center most likely used in the siting of Venus as the Morning Star because of its position facing due southeast. Like the southeast section of Knossos, this building had a spacious, open-air quality about it with its numerous windows and large rooms, one which contains an altar. The building itself was elevated by a staircase which is located on the northwest side of the temple in order to enter into the Northwest Temple of the Cult Centre. Two frescoes, one of donkeys in processional and the other of two goddesses and a god with a body in the form of a shield were found in the Southeast Temple indicating its sacral purpose. The shield in the shape of a figure eight, possibly symbolizing the eight years of the Venus cycle in conjunction with 99 moons, is seen in the frescoes of shields in the Southeast section of the temple-complex at Knossos as well.

To enter the Northwest Temple, a staircase leads steeply downward to two buildings called The House of Idols and the House with the Frescoes. These buildings, like those at the northwest section of Ħaġar Qim, are constructed as closed, private, cave-like chambers of worship characterizing them as typical for the rituals of the goddess as the Evening Star. The House of the Idols contains a vestibule, a room with a platform or raised dias, three bases for tree columns, and steps leading to a room containing and altar with eight large terracotta idols of goddesses in the *psi posture* facing the wall and eight smaller terracotta figures. To the left of the dias is an alcove in the shape of an isosceles triangle with a window that faces northwest for the viewing of Venus and an opening to the dias room. This alcove is similar to the apse and shaft chambers in the Paleolithic and Neolithic temples in its construction and location and might have been used as an oracle chamber. Several small ivory figurines, a cowrie shell or triton shell, and beads of amber and lapis lazuli as well as terracotta snakes were also found in the House of the Idols. A passageway leads to the House of the Frecoes which also served as a shrine containing bases for wooden columns, a larnax, a hearth in the center, and several goddess idols in the *psi posture*. In the House of the Frescoes is a fresco depicting two goddesses positioned on either side of a small male figure with a third, smaller goddess looking on from the lower level of the temple they inhabit.

This fresco demonstrates the importance of an emerging myth concerning the sharing of the *kouros* or Hyakinthos one portion of the year with the Minoan Persephone of the telluric world under the horizon and the remaining two portions of the year with the Minoan Aphrodite of the celestial realms. Like the myth of the Eleusinian Mysteries where Demeter must share her daughter Persephone between the earth and the Underworld thus determining the seasons, this myth also determines change through the movement of two goddesses between realms. With the myth of Venus rising from the sea in a continuous cycle of birth and re-birth, the myth of Venus and Persephone sharing the *kouros* elucidates the role of the goddess of Venus and the role of the feminine in the celestial realm in a belief system that is predominantly feminine. At the Knossos temple-complex, the role of the *kouros*, Hyakinthos or pre-Adonis figure is seen on the signet rings, in the frescoes and in statues as a figure of smaller proportions, almost as if he was the object that is to be shared.

In the fresco in the Mycenaean Cult Centre, the Minoan-Mycenaean Aphrodite stands in front of the highest level of the temple with a decorated pillar behind her facing a figure that is draped in a long, dark gown. They gesture as if they are in a decision-making process about two very small male figures attached to a staff that the Aphrodite figure is holding. One male figure is painted red signifying the *kouros* in his earthly aspect, and the other is black signifying the *kouros* in his underworld aspect. The Persephone figure is holding a sword whose point is facing downward as a posture of acquiescence, and the Aphrodite figure is holding a staff, which is attached to the male figures, in a posture of more confidence and control. After all, she has more time with her Adonis in her journey through the night sky as both the Morning and Evening Star, a journey that approximates 260 days and nights in the eastern and western skies. To the left of these figures and on another wall is a portal to the Underworld in the form of a doorway decorated with circles each containing a star. Under that doorway, is a small figure of a goddess holding what appears to be two shafts of wheat or grain with an animal at her feet suggested to be a griffin. More than likely, this is the figure of Demeter approving of the bargain her daughter makes with the celestial goddess. She wears the same long, dark dress that the Persephone figure wears and a hat that is different from those of either of the goddesses above her.

In a second fresco discovered in the Southeast Temple of the Mycenae Cult Centre, the same myth is depicted where two goddesses flank a shield that holds a human figure gesturing in the same manner as if to bargain over the central figure. The figure of the shield appears to be a male whose entire body is the figure-eight shield. In the warrior culture of Mycenae that produced the mighty Agamemnon, leader of the Trojan War heroes, and developed a code of warrior ethics, the role of the *kouros* begins to take on a significance that has not been presented in the goddess culture previously. Although the shields are displayed at the Knossos temple-complex in the apartments of the goddess, the *kouros* has not become the shield. On a signet ring from the treasures of Mycenae, the *kouros* is also seen playing an active role in the iconography. On the ring, the male figure, now the same size as the two goddesses is seen pulling the sacred tree from its position on the temple column as one angry goddess looks on and another throws herself over an altar in grief. This Adonis-like figure seemingly plucks the fruit of the sacred tree on a similar signet ring from Vapheio indicating the end of his time with the celestial goddess and his beginning journey to the Underworld with Persephone (Evans 78-9). Although the myth continues into the culture of Classical Greece as does the myth of Aphrodite rising from the sea, the beginnings of cultural change toward a more patriarchal mythology, ritual and iconography are evident in the depictions of the *kouros* at Mycenae.

On one hand, the abundance of female terracotta figures as well as the frescoes and temples of the Cult Centre reflects a definite presence of the goddesses in Mycenaean ritual and belief. However, on the other hand, patriarchal changes are beginning to dominate the culture. For instance, in the treasure trove of art unearthed at Grave Circle A, several gold crowns decorated with moons and stars representing the celestial High Priestesses of the temple-complex and an important ivory statue of two goddesses sharing the affections of a boy who crawls on their laps are clear representations of the myths and beliefs of the goddess culture (Spathari 86; 93). However, the changes to the *kouros* figure and the actual fortification of the temple-complex are clear indications of an approaching age of patriarchy and war. The Lion's Gate and main entrance of the Cyclopean walled fortification serves as an abstract representation of this change. Built in the second occupational phase of the temple-complex, the Lion's Gate consists of two female lions flanking a column in a triangular slab corbelled above the gate. On top of the column are four celestial spheres, and the lion's paws are resting on two altars. The lions are symbols of the earth powers of the goddess as are the figures of the sphinx and the griffin, and the column as The Tree of Life as well as the celestial spheres all reflect the goddess and her powers; however, the abstract and realistic representation is masculine and protective merging both the ideologies of the matri-local culture with the ensuing patriarchy.

In Classical Greek art, architecture and religion, the myths of Aphrodite are also oriented around the cycles of the planet Venus in her journey through the night sky as the Morning and Evening Star. The myth of her birth and re-birth from the sea in the eastern sky begins her mythology, and her journey in the west and to the underworld with her lover continues her mythology; however, in these myths she is represented as a figure in the mythology of an emerging patriarchy as part of its value system. Aphrodite is one of the goddesses included in the cosmogony of original creation as a heavenly representation of the feminine because she is a representative of resurrection and renewal. In a culture emerging from a dormant or Dark Age after the decline of the Bronze Age of the Mycenaean kings and rulers, the newly forming democracy based on the reforms of Solon, Cleisthenes, Theagenes, and Cylon, is bent on creating a body of knowledge in the form of myth, ritual, and philosophy that inters a belief system that includes a freedom from oligarchy and tyranny in the form of a patriarchy and democratic governing body. In such a political and religious climate, the economic and ideological rivalry between aristocratic clans and *genos* are replaced by a newly emerging value system that supports a governing body of men who are free from tyranny and chose their leaders as archons or sons of a patriarchy of newly formed values.

The inclusion of the transformative character of the archetype of the feminine as a symbol of renewal, re-birth, and change is essential to the creation of a new cosmic and earthly order represented in the original creation myths of the Classical Greeks. Hera, a representative of the precessional cycles of the moon in conjunction with the solar year and the celestial patterns of the constellations, is given the powerful position as the wife of the patriarchy to rule the seasonal and daily order of this newly formed cosmos, while Aphrodite is given the powerful position as representative of the underlying elements that are necessary to inspire change and flux as well as regeneration and renewal, all seen as elements of Venus' journey in the night sky. Moreover, Aphrodite as a symbol of femininity and love balances a culture that is immersed in masculinity and war, a fitting antithetical figure in an age that confronts change in the form of violence. Her association with resurrection and change as well as her feminine attributes and her continued worship from the Paleolithic to the age of Classical Greece make her a powerful figure for change. Perhaps this is why she is included before Hera and the creation of the Olympians and is made part of the struggle to establish the patriarchy from its cosmic roots to the current era.

In the Pelasgian, Homeric, and Orphic Creation myths, the symbols associated with the Regeneratrix, such as the waters of birth and re-birth, the dove and the waterbird, and the planetary powers foreshadow the arrival of Aphrodite in the Olympian Creation myth. In the *Genealogiae*, Hyginus establishes the creation of the Oceanides from the union of Ocean and Tethys who rule over the planet Venus. Eurynome, an Oceanide and a goddess of the powers of creation from the primordial sea, is mentioned by Pausanias as being worshipped by the Pelasgians when he visits a sanctuary of Eurynome near the sacred waters of Neda (8.41.6). Much like Aphrodite, herself, rising from the sea, this creation goddess assumes the form of a dove and lays the egg of the Universe when she couples with Ophion, the great serpent (Apollonius Rhodius 1.496). In the Homeric Creation myth, the goddess Night lays a silver egg in the womb of Darkness and sets the Universe in motion with the birth of Eros, Aphrodite's later son and companion in the Olympian Pantheon. Here, too, Tethys is the mother of creation when she mates with Oceanus (Hesiod 116; 134). The recurrent imagery of the goddess' association with water, waterbirds, the sea and her role as creatrix is transferred to the Olympian myth with the overlay of the father as the god, Uranus. Here, the planetary powers of the creation of the universe are established as masculine and dominating through the figure of Uranus.

The two versions of Aphrodite's birth, one where she is born from the masculine parts of the castrated Uranus and the other where she is the daughter of Dione and Zeus, indicate a movement to an age of patriarchy with a heavy reliance on the original forces of creation and cosmic birth still evident as vital to the process. In Hesiod's version, Cronos hates his "lusty sire," Uranus, because of Uranus' tyrant nature and usurps him, in turn, becoming a tyrant himself. Cronus' son, Zeus, usurps his father Cronus to establish the Olympians, a pantheon of justice and order. Other than establishing a forthright and sound democratic patriarchy over tyranny, the birth of the only Olympian, Aphrodite, from the original forces of creation illustrates the need to balance the creation of a new order with feminine powers that exceed the powers of the pantheon. In Hesiod's *Theogony*, Aphrodite is born from the sea foam of the castrated genitals of Uranus, and she comes ashore to Cythera and Cyprus to establish her worship centers then rises into the heavens to sit with the assembly of the gods accompanied by Eros and Desire (134-209). Her powers, as daughter of the cosmic forces accompanied by love and desire, are greater than those of all the other deities with the exception of the three virgins, Athena, Artemis and Hestia, whose interests do not include the ways of love and marriage (*Homeric Hymns to Aphrodite V*: 1-44).

In hymns to Aphrodite, Homer praises her power over the gods and goddesses and establishes her overwhelming beauty as golden and bright *Homeric Hymns VI and X*). Aphrodite is welcomed by the Hours, symbols of time and cosmic order:

> They clothed her with heavenly garments: on her head they put a fine, well-wrought crown of gold, and in her pierced ears they hung ornaments of ori-chalc and precious gold, and adorned her with golden necklaces over her soft neck and snow-white breasts, jewels which the gold-filleted Hours wear them-selves whenever they go to their father's house to join the lovely dances of the gods. (*Hymn VI*: 6-11).

The epithet of Ourania or Heavenly is ascribed to Aphrodite because of her association with Uranus and her sanctuaries are established in Cythera and at Corinth in honor of her place among the Olympians: "In addition to these major centers of worship, many smaller-scale cult sites throughout Greece, from Argos to Thebes, from Athens to Elis, were dedicated to Aphrodite Ourania" (Rosenzweig 60). As Aphrodite Ourania, a major altar is also established to her in the Agora at Athens, and she becomes a symbol of unity and cosmic order to the Athenians.

The imagery associated with the goddess of Venus, such as her cosmic powers, her epithets as golden and bright, and the aforementioned associations with the sea and waterbirds, is continued in another version of her birth cited by Apollodorus in *The Library*, a second century compilation of myth, epic, and tragedy. Apollodorus states that Aphrodite is the daughter of Zeus and Dione, a daughter of the Titanides, symbols of original creation (1.3.1). Dione, like Aphrodite, is a goddess of cosmic power and oracular vision. Her name is the feminine version of Zeus intimating that he is a symbol for creation borrowed from the goddess. Moreover, Dione has an oracular shrine, much like the oracle chambers of the Venus temples in previous cultures, which Zeus takes over at Dodona. Dione's priestesses collectively known as the Peleiades, are three old women named "Doves" or *peliai* who guard the shrine of Aphrodite in their temple (Strabo *Geography* 7.7.12). Dione, too, is a daughter of Oceanus and Tethys, two cosmic forces whose planetary power is Venus. Aphrodite's lineage, in each version of her birth, is attuned to her history and development as goddess of regeneration and cyclical birth through her travels in the night sky. Her birth from the sea, like the many images cited from the Paleolithic to the Classical Age, marks her importance as the Eastern Star rising to create and re-create the cycle of resurrection and birth.

The Acrocorinth, a rocky acropolis overlooking the *Korinthiakós Kólpos*, is similar to many of the Neolithic and Bronze Age temples devoted to the Regeneratrix as Venus which are oriented to view Venus rising from the sea. Scanty remains of a sanctuary to Aphrodite were found on the northwest side of the Acrocorinth, and on the north side of the Acrocorinth, a long rampart commences to the sea and to another sanctuary dedicated to Aphrodite and Poseidon. Most likely, this temple and rampart was used for ritual processions to descend to the sea where the suppliants might bathe in the waters as a symbol of re-birth and renewal. The attachment of the birth of Aphrodite rising from the sea and her associations with water as a symbol for resurrection and re-birth is also evident in the 6th and 5th century Agora and Temples of the ancient town of Corinth located at the base of the Acrocorinth. At the base of the Acrocorinth, the sacred waters from the foothills of the acropolis were channeled via a conduit from their sacred source to the Glauke fountain, the Lerna Fountain, and into the baths of the Asklepieion for healing purposes. In the southeast section of this temple-complex is a Temple to Aphrodite dedicated to her as a goddess of the healing waters. Among the art discovered in the temple-complex is a mosaic floor with patterns of griffins, terracotta doves, nymphe statues, and a terracotta head of a cult statue of Aphrodite.

At many of the Mycenaean and Archaic temple-complexes of Greece, such as the cult centers at Epidaurus and Ilissos, Aphrodite's temple is dedicated to the rejuvenating powers of the goddess represented in the creation myths of Aphrodite that are associated with water and rebirth. At Epidaurus, the sanctuary of the Aphroditeion, a small building of porous stone, is located near the original entrance to the temple-complex near a sacred well. Myrtle trees still grow in this once lustrous meadow that provided sacred waters for healing. Later, in the 5th century BC, the Baths of Asclepios were added, and then, the Stoa and Roman Baths were added to this part of the temple-complex in much the same manner similar structures were added at Corinth to the original structures of the Mycenaean and earlier ages. A similar temple dedicated to Aphrodite stands on the banks of the Ilissos River in a temple-complex southeast of the Athenian Acropolis. The location and proximity to the river indicate that is may have been used as a temple of rejuvenation and re-birth. According to Pausanias, the temple to Aphrodite at Ilissos was in a "district called The Gardens" where a statue to Aphrodite stood near the temple with a "square shape like that of the Hermae, and the inscription declares that the Heavenly Aphrodite is the oldest of those called Fates" (1.19.2). Pausanias' reference to the statue indicates its archaic structure as a herm or plank, like the sacred *xoana* which were ritually bathed, and his reference mentions Aphrodite's connection to the heavens as Aphrodite Ourania.

In the myths of Venus as the Evening Star, Aphrodite enters into complex relationships with several suitors, both immortal and mortal, which emphasize her new role in the patriarchy as well as her vigilance as a goddess of fertility, procreation, and sacred love. The myth of the goddess of Venus as Evening Star with the *kouros* or Hykanthios of Minoan-Mycenaean symbology is continued in the myths of Aphrodite with Anchises and Adonis as her *kouros*. Both of these myths include incestuous desires for a daughter by a father and act almost as moral lessons where the powers of Aphrodite must be curtailed in some way in a culture that is establishing the rules of paternity. Zeus himself is tempted to lie with his daughter, Aphrodite, so he attempts to humiliate her by having her fall in love with a mortal, Anchises. In *Homeric Hymn V*, Aphrodite retains many of the qualities of the Minoan-Mycenaean goddess of the Evening Star. She comes to Anchises' bed from her fragrant altar where she has bathed in a lustral basin anointed with sacred oils by her attendants, the Graces, and "decked with gold," and she is followed by wild beasts who mate, "two-together, about the shadowy coombes" as they approach Mt. Ida in the Trojan countryside (53-74). Reminiscent of the Lion Gate at Mycenae where two wild beasts guard the temple-complex of the goddess, Aphrodite approaches her bed with Anchises accompanied by paired beasts "like a pure maiden in height and mien" (80).

When Anchises discovers that this is no ordinary nuptial, he fears the retribution of the gods, but Aphrodite assures him he will live to a ripe old age, and she will bear the humiliation of lying with a mortal thus curtailing her powers with the Olympian deities. She says she will return after her four year cycle in the heavens and bring him a son, Aeneas, who will become the head of a royal lineage and hero in the Trojan War. Anchises, Olympian Aphrodite's first *kouros*, is more of a model for the establishment of patriarchal lineage than he is a dying god because his myth represents a renewal of life through his son who returns after the four year cycle of Venus in the heavens, and he lives a long, full life. The myth of Aphrodite's love with Adonis, on the other hand, is a continuation of the myth depicted in the fresco at the Mycenae Cult Centre and in the myths of the goddess as Venus, the Evening Star because of Adonis' untimely death. Again, the patriarchal taboo of incestuous love between a father and a daughter begins the myth of Adonis' birth where Aphrodite must rescue Adonis from his mother, Smyrna, who has been turned into a myrrh-tree as punishment for incest with her father. Adonis is born from the tree ten months after Smyrna has been transformed, and Aphrodite conceals him in a chest to be guarded by Persephone in the Underworld where Persephone has been instructed not to look into the chest (Apollodorus 3.14.4).

Again, the myth contains many of the symbols of previous myths of the goddess of Venus with the addition of images used in the Classical era thus creating an Aphrodite that represents both the objective correlatives of previous cultures that aimed at tying celestial events with the sacred and divine as well as the concepts of a newly formed patriarchy. The Adonis myth retains the idea of the goddess of Venus climbing or mounting The World Tree or The Columns of Life in the symbolic myrrh-tree that bears Adonis, and it also contains the amount of time for Adonis' birth as the young *kouros* as the time Venus is again seen in the west as the Evening Star in her 260 night cycle. Moreover, when Persephone opens the chest and also falls in love with Adonis, an Olympian court decides that Adonis will spend two thirds of the year with Aphrodite and the other third with Persephone clearly outlining the entire cycle of Venus in the night sky. Adonis' eventual death by a boar, attributed to Ares' jealousy for any lover Aphrodite is truly in love with, is the culmination of Venus' journey in the Underworld where her *kouros* is now dead and must be retrieved from death and his time with Persephone. This time marked on the Attic calendar of the Greeks when Venus disappears below the horizon, contains events that culminate the journey of Venus in the night sky in festivals called the Adonia, the Arrephoria and the Aphrodisia.

Neolithic calendars that were adapted from the megalithic structures, such as Stonehenge and Carnac, served as a basis for the Bronze Age peoples who transferred the information of luni-solar, planetary, and stellar calendar cycles to bronze tablets or mechanisms such as the Coligny Calendar from Coligny, France and the Anti-Kythera Device from Greece, respectively. The translation of this information led to the sophisticated calendars of the Druids in Europe and the Attic Calendar of the Greeks which accurately measured the cycles of precession and charted their festivals and religious observances accordingly. The cycle of Venus, in conjunction with the moon, the sun, and the stars is recorded on the Coligny Calendar and on the Attic Calendar in terms of four year cycles where Venus is measured in the same time and place in the night sky in conjunction with the moon and the sun. When Venus goes through three cycles of 260 nights as the Morning Star on the eastern horizon, three cycles of 260 nights as the Evening Star on the western horizon, and approximately 60 nights under the horizon, at the end of the third cycle, the planet returns to the same moons for its final phase of 60 nights under the horizon as it did four years ago thus determining the four year cycle of Venus as sacred on the ancient calendars. In the myths of the goddess as Regeneratrix, her Eastern Star journey represents her cosmic birth and re-birth, her Evening Star journey represents epiphany of love with her *kouros*, and her journey under the horizon represents the loss of her lover to the Underworld.

In the fourth year cycle of Venus, this return to the Underworld phase is marked by religious festivals on both calendar systems as the beginning of a precarious and sacred cycle. Perhaps the ancients marked these nights when Venus is under the horizon as a particular time to pray for the return of the goddess when she is eventually spiritually and physically resurrected from the Underworld, cleansed and renewed from her journey of retrieving Adonis. On the Celtic calendar of the Druids, this cycle begins at the end of Cutios, the sixth moon, and continues through Giamonios and the Simivisonnios, the seventh and eighth moons. On the Attic Calendar, this cycle begins at the end of Thargelion, the sixth moon, and continues through Skiraphorion and Hekatombaion, the seventh and eighth moons. For the Classical Greeks, the festival at the end of Thargelion and the beginning of Skiraphorion contains the solar holiday of the Summer Solstice or the declining of the light of the sun and the beginning of the dark half of the year; it is called the Adonia which mourns Aphrodite's loss of her lover to the Underworld. The festival in the month of Skiraphorion is celebrated on the full moon of Skiraphoria, and it is called the Arrephoria; again, it is a festival of descending to the Underworld. Finally, the festival in the month of Hekatombaion is the time to end the vegetative cycle of the year and the life of Adonis who is a vegetative spirit; it is called the Aphrodisia. On both the Celtic and Greek calendars the end of these festivals is a time for the renewal of life forces and the cleansing of the spirit.

The Adonia is celebrated in Sicily as the Anagogia, in Argos as the Hysteria, and in Mesopotamia as the death of Dumuzi-Tammuz. It is a festival of mourning for the death of the goddess' *kouros* and is marked as a time of loss, sacrifice and embarking. In the Adonia, women pot plants with short-lived maturation periods, such as fennel and lettuce, which will sprout and die quickly. These Adonis gardens are on roof tops and when the plants die, the women lament in a festival. According to Walter Burkert, "The atmosphere of the festival is infused with the sweet aromas of incense, but the climax is loud lamentation for the dead god. The dead Adonis was then laid out on his bier in the form of a statuette and borne to his grave: the effigy and the little garden were thrown into the sea" (177). The demise of Adonis into the sea culminates the meaning of the myth where Aphrodite sends him to Persephone in a chest in the sea, and it marks the descending of Venus below the horizon as the end of her journey as the Evening Star. A ladder was used in the ceremony to get to the roof tops, and this same ladder is seen in the iconography of Aphrodite at the Adonia where she descends the ladder for Adonis' demise. The ladder appears on vase paintings, hydria, on the lebes gamikos with Aphrodite and her attendants (Rosenzweig 65). A ladder to Aphrodite stands in the courtyard of the Pandroseion or the open-air garden of the Erechtheion on the Acropolis, and many votive ladders to Aphrodite were discovered in the Sanctuary of Nymphe now displayed in the Acropolis Museum.

The second festival to Aphrodite, the Arrephoria, is similar to the Adonia because it ritualizes Aphrodite's descent to the Underworld and her mourning for the dead Adonis. It takes place every four years at the Acropolis Pandroseion which is behind the Porch of Maidens of the Erechtheion where the sacred re-planted olive tree and the ladder are found. According to Pausanias, maidens are appointed by the Archon Basileus of Athens as priestesses of Aphrodite in preparation to participate as Arrephoroi or "Bearers of the Sacred Offerings" in a ritual where they carry unknown objects on their heads down a "natural underground passageway" or Mycenaean stairway from the Erechtheion to the Gardens of Aphrodite located at the northwest slope of the Acropolis. The Arrephoroi "leave down below what they carry and receive something else which they bring back covered up" never knowing the contents of either container (*Descriptions of Greece* 1.27.3). In a myth about three Arrephoroi named Aglauros, Herse, and Pandrosos, Pandrosos was the only one who didn't look at the objects; she was given the honor of attending the sacred olive tree, and the other two Arrephoroi jumped to their death off the steep cliffs of the Acropolis in shame. In these accounts, the symbol of the olive tree, sacred to both Aphrodite and Athena, is still an important representative of The World Tree and The Columns of Life from the Paleolithic era. Moreover, the Garden of Aphrodite on the northwest slope of the Acropolis is near caves that resemble the temples to Aphrodite in the Paleolithic, Neolithic and Bronze Ages.

In the niches of the Garden of Aphrodite, phallic totems, nude male terracotta figures, and genitalia votives were placed as symbols of holding, cherishing and worshipping the fertility and sexual pleasure associated with the *kouros* as Adonis. More than likely, these were the secret objects that the Arrephoroi were not allowed to see because of their virginity. The surname of Aphrodite in association with the gardens is En Kepois in Athens and Hierokepia on Paphos. In an open-air sanctuary to Aphrodite at Daphni, a temple about seven miles from Athens, similar votives of male genitalia, doves, and terracotta figures were discovered in the niches of a rock cliff. A cult statue of Aphrodite leaning against a tree associates her with Aphrodite En Kepois or Aphrodite of the Gardens in this temple as well (Rosenzweig 40-4). Another festival in Skiraphorion, the Skiraphoria, is named for the Eleusinian hero, Skiros, who died in a battle against Erechtheus the king of Athens who also died in the same battle. Part of the Skiraphoria was a processional from Athens to Eleusis where the procession stopped at the rock cliffs of the temple at Daphni to honor a dead king and a hero. Skira, like Adonis, is a young hero who is worshipped after his heroic death and the word *skiros* itself means "something like white earth" (Burkert 230). Perhaps *skiros* is referring to the pallor of death of the young hero and/or the white sculpture votives that were placed in the niches of Aphrodite of the Gardens at Daphni.

The last festival on the Attic calendar dedicated to Aphrodite's journey to the Underworld is the Aphrodisia. In the Aphrodisia, a ceremony was performed where a dove was sacrificed on Aphrodite's altar, her altar was then anointed and purified, and her cult images were conducted in procession to a place where they were washed (Simon 48-9). The Aphrodisia is a fitting conclusion for the preparation of the end of Aphrodite's journey from the Underworld to her rising as the Eastern Star. Here, Aphrodite is prepared for resurrection and the beginning of another cycle in the night sky. A sanctuary for the Aphrodisia was built between the Asklepieion and the Propylaia as a dedication to Aphrodite Pandemos during the Hellenistic era where the surname of Pandemos indicates the goddess as a force of unification of the people or *demos*. Two cult statues, one of Peitho, a companion of Aphrodite's who also represents democracy and the art of argument and persuasion admired by the democracy, and one of Aphrodite, were the cult statues that were in the ceremonies. The inscription on the aedicule of Aphrodite Pandemos indicates that the ritual of washing the statues is honoring an ancient ritual most likely referring to the washing of the *xoana* in the Minoan-Mycenaean rituals of Aphrodite and Hera. It also reiterates the idea of the washing of the bride in epaulia scenes on *lebes gamikos* and *pyxis* before marriage where Aphrodite Pandemos and Peitho are depicted overseeing the rituals in order to stabilize marriage as a harmonizing force of Athenian democracy (Rosenzweig 21-2).

On a physical level, the Aphrodisia represents the unification of the polis in its efforts to portray a democratic body politic, and on a spiritual level, the festival represents the resurrection of the goddess as the goddess who once again is cleansed and ascends to the heavens. Framing the aedicule of Aphrodite Pandemos are six doves "carrying in their beaks knotted fillets like those hung around the neck of sacrificial animals" (Rosenzweig 14). The sacrifice of the doves may be seen as the transference of power needed to give Aphrodite her wings to ascend as her bird attribute or soul or it may be seen as a sacrifice made to the goddess for a successful ascension to the heavens. The altar of the Demos and the Graces found in the Agora and the marble doves from the Daphni shrine of Aphrodite among other works of art were most likely used for similar ceremonies as those in Attica at the Aphrodisia. Aphrodite's association with doves and waterbirds continues as an apt image of soulful flight and transition or epiphany seen on statues and vases throughout the Classical era, one of the most notable being Aphrodite riding on a swan on a *kylix* by the Pistoxenos Painter. The Classical artists who represent Aphrodite with her bird totem or in epiphany form add the context of political peace, unification and the corresponding unification of the sexes in marriage as part of the worship of Aphrodite in the Greek democracy.

In the myths of Aphrodite associated with the emerging patriarchy and, most notably, the myths associated with the Trojan War, Aphrodite still holds a position of power and influence as a force of change and as a symbol of the antithesis of war and violence; however she is often compromised because of her amorous nature. In Homer's *Odyssey*, a myth of Aphrodite's affair with Ares, the god of war, establishes Aphrodite as the antithetical symbol to war. Her relationship with Ares, according to Burkert, is "developed more as a polarity, in accordance with the biological-psychological rhythm which links male fighting and sexuality, thus the *polemarchoi* of Thebes sacrifice to Aphrodite at the end of their term. The daughter of Ares and Aphrodite is Harmonia. Joining, which at the same time denotes musical euphony, sprung from the conflict of war and love" (220). In *The Odyssey*, Odysseus' young crewmen dance to the lyre and play a ball game with a similar "beautiful purple ball" that the Arrephoroi use in their games on the Erechtheion seen on vase paintings (Rosenzweig 50). Their musical celebration is devoted to the release of Aphrodite and Ares from their adulterous love affair where Hephaestus, Aphrodite's husband in the Olympian Pantheon, releases them from the bonds of blame both physically and spiritually most likely because of the necessary and natural bonds that love and war share.

Hephaestus frees Aphrodite and Ares from the bonds he has forged from his great anvil, when he has caught them early in the morning in bed together. Helios, the god of the sun, was the first to observe Aphrodite and Ares together, thus setting in motion Hephaestus' successful scheme to capture them. When Aphrodite is freed and absolved, after much consternation among the gods that Hephaestus has called together to witness the love scene, Aphrodite goes to Paphos on Cyprus to her temple "where is her demesne and fragrant altar. There the Graces bathed her and anointed her with immortal oil, such as gleams upon the gods that are forever. And they clothed her in lovely raiment, a wonder to behold" (Book VIII: 363-66).

Aphrodite is absolved and restored to her former status, and the listeners to the tale rejoice and continue their dance and ball games celebrating like Arrephoroi in the presence of their goddess. The myth occurs when Venus is the Eastern Star, and the celebration of bathing in the sea as Venus rises from the sea is echoed in Aphrodite's ritual of renewal at Paphos. The inclusion of Helios in the myth most likely refers to Venus passing in front of the sun every nine months in the beginning of her journey as the Eastern Star. In both cases, Aphrodite's ability to renew her splendor and power is restored and the myth of Aphrodite as the Eastern Star is recounted as a tale to assuage the heroes of war.

Where the myth of Aphrodite as the Eastern Star is used in the aftermath of the Trojan War, the myth of Aphrodite as the Western Star is used in the onset of the saga of the Trojan War. In the myth of The Judgment of Paris summarized by Proclus from Stasinus' *Cypria*, Aphrodite wins a contest judged by Paris based in which one of three goddesses, Aphrodite, Athena, or Hera is *kallistē* or most beautiful (*The Epic Cycle* I: 1-11). Zeus and Themis already plan the Trojan War as a way, perhaps, of setting in motion the arena of demi-gods and heroes in the course of human events. They omit the goddess of strife, Eris, from the wedding of Peleus and Thetis, and when Eris tosses a golden apple to the goddess who is most beautiful and the three goddesses argue over who is most worthy of the apple, Zeus chooses Paris, son of Priam, King of Troy, to judge who will be the winner. Aphrodite offers Paris the love of Helen of Sparta as his prize if she is chosen, and she wins over the gift Athena offers which is wisdom and the gift Hera offers which is power and land. Aphrodite as a representative of the power of love and beauty wins the golden apple which complements her golden attribute. In this myth of love and war, again, Aphrodite is seen as the antithetical statement of love and renewal as well as the procreative power of resurrection necessary to balance the strife of war.

Moreover, the myth of the love story of Helen and Paris is a continuation of the myth of Aphrodite as the Western Star where the goddess, or in this case the demi-goddess, is united with the *kouros,* or in this case the hero, as her lover. As in the case of the Greek Creation myths, the myths of the Age of Heroes reiterate the earlier Bronze Age myths of Aphrodite changing them to fit the advent of warfare and the creation of demi-goddesses and gods; evidence of Helen as a Proto-Indo-European sun-maiden connects her to myths concerning her suitors and her marriage to powerful figures (Dexter 41). In the Classical era, Helen of Sparta serves as the representative of Aphrodite in the war saga. Helen is the demi-goddess who is connected to Aphrodite in the Trojan War because she has the attributes of Aphrodite such as her association with waterbirds and the flight of the soul ascending the heights of the spiritual realm; her abilities of renewal and rebirth from the sacred waters; and, her cosmic love for the *kouros.* From her birth, Helen is a deity associated with waterbirds. Helen was born from Leda and Zeus, in the form of a swan, or from Nemesis and Zeus who were in the forms of a goose and a swan, respectively. In the version of Helen's birth myth, Nemesis still gives the egg she bears to Leda and Leda raises the demi-goddesses, Helen and Clytemnestra, who shape the Age of Heroes in mythology (Apollodorus 3.10.7). Like Aphrodite, Helen's lineage is linked to the goddess as Regeneratrix through her association with the immortal power of the bird as a symbol of the ascension of the soul, and like Aphrodite, Helen's birth is part of the creation myths of a culture that is shaping a new ideology.

Helen is also associated with the sacred waters of renewal and rebirth near those of Aphrodite. Near Corinth is the tiny village of Loutra Elenis, Helen's Baths, a ritual site for therapeutic waters and renewal of spiritual energy; the site, according to Bettany Hughes in her book *Helen of Troy* is still an invigorating spring that enters into the sea (251-52). Likewise, a site for Helen's temple on the hill of Therapne is located near a spring described by Pausanias as such: "On the road that leads through the mountains to Hermione is a spring of the river Hyllicus, near the rock is a sanctuary of Aphrodite Nymphia made by Theseus when he took Helen to wife" (2.32.7). The landscape and especially the sacred waters are marked with the history of both Helen, whose daughter was named Hermione, and Aphrodite, whose sanctuary is close by. Most importantly, is the association of Helen with a *kouros* figure in mythology. In mythology and in art, Paris is depicted as a beautiful, youthful figure. In *The Iliad* when Paris comes forth to do battle with Menelaus, he is described as "champion godlike Alexander, bearing upon his shoulders a panther skin and his curved bow, and his sword" (3.17). Even in the words of his enemy, Hector, who chides Paris for his adulterous behavior, Paris is "most fair to look upon" with a "comely form" and long "locks" (3.38; 3.54). In a bronze sculpture of Paris which was part of a treasure trove brought up from a shipwreck off the coast of Antikythera, Paris' figure is that of a young and handsome man, a *kouros* of the earthly representative of Aphrodite: Helen of Sparta.

Of all the depictions of Aphrodite in the Classical Age that brings forth her journey in the night sky as the Morning and Evening Star and as the goddess of regeneration, renewal and resurrected love is what might possibly be the most splendid effort in adoration: her altar from Lokroi Epizefiri in Sicily. In the caves of the Paleolithic temples to the goddess as Regeneratrix, the altars of the stalactites and stalagmites that meet to form The Columns of Life with icons at their base were the first altars to the deity. In the Neolithic era, these altars became megalithic trilithon structures with the symbols of the goddess intact. An offering table with a small cup-like indentation was found at the top of Mt. Ida near the Knossos temple-complex in Crete, and an altar to Aphrodite Hegemone and the Graces in the Agora as well as an altar to Aphrodite Pandemos at the Acropolis were part of her worship in Athens. However, an altar created in an area peripheral to mainland Greece depicting Aphrodite rising from the sea with the aid of two attendant priestesses or Nymphes, has been paired with a second sculpture portraying the myth of Aphrodite, Persephone and Adonis which was discovered in the same region of Rome where the Sicilian altar piece was transported after the Romans conquered Sicily in the first century BC. The two sculptures fit together, and thematically they complete the myths of Aphrodite in the Classical Age.

Both sculptures are shaped like the headboard and footboard of a large bed. The reliefs of the myth of Aphrodite as the Morning and Evening Star face outwards as the headboard and footboard, while the shorter ends that depict adorants or worshippers on either side of the main scene act as the sides of bed. If the footboard is the scene where Aphrodite rises from the sea in the eastern sky and the headboard is the scene where Aphrodite is in the western sky, the matching sides correspond in theme and are easily paired. On one side are two young adorants; one is a nude female playing a flute with double pipes and the other is a nude male playing a lyre. On the other side are two elderly adorants; one is a woman with a thymiaterion and the other is seemingly a male holding what must have been a similar offering. The two young nudes and Aphrodite who is partially clothed herself all sit on cushions with highly suggestive folds that resemble the nude pelvic area of a woman. On the familiar scene of Aphrodite being born from the sea, the two Nymphes or priestesses assist Aphrodite with a play of rising curves and seemingly wet garments that show the top half of the goddess' body nude and the bottom half-hidden by a drapery. On the other relief, Eros holds what was once a three-dimensional scale where two small *kouros* are dangled from the scale hanging and bound by their hands over their heads. The scales are tipped to the side where Aphrodite sits on her erotic cushion raising her hand in a triumphant position. On the other side of Eros' scales, Persephone sits with her head bowed in sadness, her hand holding her head.

The dramatic image created by the sculpture incorporates the mythology of the Venus into one coherent form where astronomy, art, archeology and myth combine to tell the story of the Regeneratrix resurrecting from her watery depths in a never-ending cycle of time. Together with the cycles of the sun and the moon, the cycles of Venus reflect what T.S. Eliot refers to as the "objective correlative" necessary for human consciousness to understand the divine. The bed-like altar to the goddess is therefore an apt metaphor for the transcendent experience that the goddess brings to humanity in the form of the belief in the continuing cycles of life and death. On each side of the altar, the myths and rituals of the goddess clearly indicate the meaning of her worship. By portraying the cycles of Venus in the night sky, the myths of the goddess come alive with meaning by connecting humanity to the cosmos and the divine.

WORKS CITED

Apollodorus. *The Library*, Trans. Sir James George Frazer. *Theoi Greek Mythology: Exploring Mythology in Classical Literature and Art*. New Zealand: The Theoi Project, Ed. Aaron Atsma, 2008. http://www.theoi.com/Text/Apollodorus1.html.

Apollonius Rhodius. *Argonautica*, Trans. R.C. Seaton. *Theoi Greek Mythology: Exploring Mythology in Classical Literature and Art*. New Zealand: The Theoi Project, Ed. Aaron Atsma, 2008. http://www.theoi.com/Text/ApolloniusRhodius1.html.

Burkert, Walter. *Greek Religion*. Cambridge: Harvard University Press, 2003.

Dexter, Miriam Robbins. *Whence the Goddesses: A Source Book*. New York: Teachers College Press, 1990.

Evans, Arthur J. *The Mycenaean Tree and Pillar Cult and Its Mediterranean Relations*. London: Macmillan, 1901.

Gimbutas, Marija. *The Living Goddesses*. Edited and Supplemented by Miriam Robbins Dexter. Berkeley: University of California Press, 1999.

Hesiod. *Theogony*, Trans. H.G. Evelyn-White. *Theoi Greek Mythology: Exploring Mythology in Classical Literature and Art*. New Zealand: The Theoi Project, Ed. Aaron Atsma, 2008. http://www.theoi.com/Text/HesiodTheogony.html.

Homer. "Hymns to Aphrodite V, VI, X" *Homeric Hymns*, Trans. H. G. Evelyn-White. *Theoi Greek Mythology: Exploring Mythology in Classical Literature and Art*. New Zealand: The Theoi Project, Ed. Aaron Atsma, 2008. http://www.theoi.com/Text/HomericHymns3.html#6.

———. *The Odyssey*, Trans. A.T. Murray. *Theoi Greek Mythology: Exploring Mythology in Classical Literature and Art*. New Zealand: The Theoi Project, Ed. Aaron Atsma, 2008. http://theoi.com/Text/HomerOdyssey8.html.

———. *The Iliad*, Trans. A.T. Murray. *Theoi Greek Mythology: Exploring Mythology in Classical Literature and Art*. New Zealand: The Theoi Project, Ed. Aaron Atsma, 2008. http://www.theoi.com/Text/HomerIliad3.html.

Hyginus. *Genealogiae*, Trans. Mary Grant. *Theoi Greek Mythology: Exploring Mythology in Classical Literature and Art.* New Zealand: The Theoi Project, Ed. Aaron Atsma, 2008. http://www.theoi.com/Text/HyginusFabulae4/html.

Hughes, Bettany. *Helen of Troy*. New York: Knopf, 2005.

Nilsson, Martin P. *The Minoan-Mycenaean Religion and Its Survival in Greek Religion*. New York: Biblo and Tannen, 1928.

Pausanias. *Descriptions of Greece*, Trans. W.H.S. Jones. *Theoi Greek Mythology: Exploring Mythology in Classical Literature and Art.* New Zealand: The Theoi Project, Ed. Aaron Atsma, 2008.

Proclus. *The Epic Cycle*, Ed. Gregory Nagy. *The Perseus Digital Library,* Ed. Gregory Crane. Boston: Tufts University, 2005. http://www.stoa.org/hopper/text.jsp?doc=Stoa:text:2003.01.004.

Simon, Erika. *Festivals of Attica: An Archeological Commentary*. Madison: University of Wisconsin Press, 1983.

Spathari, Elsie. *Mycenae: A Guide to the History and Archeology.* Athens: Hesperos Editions, 2004.

Strabo. *Geography*, Trans. Horace Leonard Jones. *Theoi Greek Mythology: Exploring Mythology in Classical Literature and Art.* New Zealand: The Theoi Project, Ed. Aaron Atsma, 2008. http://www.theoi.com/Titan/TitanisDione.html.

Rosenzweig, Rachel. *Worshipping Aphrodite*. Ann Arbor: University of Michigan Press, 2007.

Chapter Three

Love Goddesses of the Early Historic Age

Miriam Robbins Dexter

INTRODUCTION

In this chapter, we look at several love goddesses, beginning with "Venus"-figures of the earliest Upper Palaeolithic and Neolithic, and then we concentrate upon the goddesses named in early historic texts. Powerful multi-functional "Great"-Goddesses, one of whose functions oversees the realm of love, are associated with the planet Venus and with 'wandering' in search of their lost loves.

UPPER PALAEOLITHIC FEMALE FIGURES

Elsewhere in this book the famous Upper Palaeolithic "Venuses" of Laussel, Willendorf, and others are discussed. Here, we would just draw attention to two figures. The first, a figure just discovered in fall, 2008, is a tiny nude female figurine; she was found in fragments on the floor of the Hohle Fels cave in southwestern Germany, near Ulm and the Danube headwaters. This is the oldest Upper Palaeolithic female figurine ("Venus figure") yet found, dating to between 33,000 and 38,000 BCE.[1] The figurine was found in a cave in which other ritual artifacts, such as a flute dating to the same Palaeolithic era,[2] were also found. Thus, this cave served as a ritual space.

Another figure was recently highlighted in a documentary film about Chauvet Cave, a vast underground cavern in the Ardèche gorge in the south of France. The documentary, "Cave of Forgotten Dreams," directed by Werner Herzog, illustrates, among many animal figures, a figure with large pubic triangle, dating to the Aurignacian period, carved on a limestone outcropping. Some time later, a bison was superimposed upon the female figure.[3] She is found in the last and deepest chamber of the cave, the "Salle du Fond." She may well have an apotropaic function,[4] guarding the entrance to the gallery.

NEOLITHIC FEMALE FIGURES

Göbekli Tepe and Sacred Display

The quality of apotropaia which one finds in Chauvet Cave continues in the Neolithic. A particular type of dancing figure begins to be found in the earliest Neolithic: a figure who not only does a magical dance but who also does an apotropaic display of her genitalia: that is, her genitals were sacred and protective. Probably the earliest (uncontested) female display figure, dating to the aceramic, pre-agricultural Neolithic no later than 8000 BCE, has been found in level II of the southeast Anatolian site of Göbekli Tepe; she is nude and crouching, and her arms and legs are bent, in an M-position, probably in a magical dance. She was carved on a stone slab which was found in the Lion Pillar building on the floor at the entrance to the room, between pillars containing depictions of felines, foreshadowing female figures associated with felines throughout Anatolia, the Indus Valley, and historic India.[5] One had to step on her in order to enter the room; she was therefore the ground upon which one walked, and she was very likely apotropaic:[6] that is, she was the goddess who protected the worshipers in the temple. This apotropaic function has been assigned to female display figures for millennia; the early historic Greco-Roman Medusa and the medieval era Irish Sheela na gigs are examples of historic female figures whose functions included the apotropaic.

UPPER PALAEOLITHIC AND NEOLITHIC FEMALE MAGICAL DANCERS

Magical female dancing and display figures have a long history, dating as far back as the Upper Palaeolithic, and they are found throughout Eurasia. Many are depicted nude — similarly to love goddesses — because the female body was considered sacred and propitious,[7] and the nude dancer was thus able to effect the magic needed for the group. Some of these dancing figures may represent prehistoric shamans, who mediated the worlds of the seen and the unseen for other members of their groups. Cave paintings dating to the Upper Palaeolithic (ca. 36,000 BCE - 15,000 BCE) depict figures doing what appears to be a magical dance.[8] There are many such figures dating to the Neolithic. A dancing figure from Swidwin, Poland, dates to ca. 9000 BCE.[9]

These dancers are depicted in Neolithic rock art[10] as well; Yosef Garfinkel, reporting on dancing scenes from Nevalı Çori, in Southeast Anatolia, discussed and illustrated prehistoric dancing scenes from Egypt to Southern Europe.[11] These dancing scenes are depicted through the prehistoric and historic ages — up through the present.

THE NEOLITHIC

Throughout the Neolithic — from at least 6500 BCE to around 3000 BCE — nude figurines were excavated from sites in Europe, especially southeast Europe, and the Near East. Whereas in the Upper Palaeolithic female figures were depicted with particularly generous proportions, by the time of the late Neolithic, ca. 3500, most Neolithic figures were slender. Some figures from this era are still depicted as large "earth-mother" types, for example figures from Malta such as the Sleeping Lady from the Hypogeum and the large female statue from Tarxien. Many early Neolithic figures, for example the woman depicted between two felines found in a bread oven in Çatalhöyük, in South Central Turkey (ca. 5500 BCE),[12] and the marble figure with polos from sixth millennium Sparta,[13] have steatopygous buttocks.[14]

LATE NEOLITHIC TO BRONZE AGE — INDUS VALLEY FIGURES

In the prehistoric (Bronze Age) Indus Valley/Sarasvatī Valley, there are two sorts of female figures which presage early historic Indic erotic goddesses. A large wood sculpture (28 inches high) of a nude female, from the Harappan culture, dated to ca. 2400 BCE, squats and displays her genitals,[15] in a manner much like early historic Indic Lajjā Gaurī figures, nude figures which hold their legs in positions which result in a bold display of their genitals; the historic figures are found in temples as a symbol of good fortune. There are also Indus/Sarasvatī Valley seals of female figures in a display position.[16] The Bactria-Margiana Complex too has produced a pre-Lajjā Gaurī figure in the form of a flower, very similarly to those produced much later in the early centuries of this era. The Indus/Sarasvatī cultures have produced seals depicting Durgā-like figures as well — female figures associated with tigers.[17]

WATERBIRDS AND SNAKES

Waterbirds were significant in the iconography of the Upper Palaeolithic. These waterbirds continued to be represented in Neolithic imagery surrounding female figures, and other birds — as well as snakes — were added as well. In the Neolithic, female figures were very often depicted as hybrids of bird/woman, snake/woman, and bird/snake/woman. These female figures represented a goddess of the life continuum: birth, death, and regeneration. The iconography of bird and snake, added to a female figure, rendered this figure a "Great"-Goddess both in prehistory and in the early historic age. This iconography is not accidental; both the bird and the snake are potent representatives of all phases of the life continuum. Birds represented two phenomena: birds such as the dove, which were associated with goddesses such as the Greek Aphrodite in the classical age, personified the breath of life and perhaps the soul. Even in modern Western cultures, the dove represents the soul, peace, and purity. On the other hand, the owl and other raptors are representations of night and death. Raptors such as the vulture and the crow, as carrion-birds, represent the battlefield and death. [18] Representations of the vulture are known from ca. 6100 BCE, from the Neolithic town of Çatalhöyük mentioned earlier. There, the excavator, James Mellaart, discovered wall paintings of vultures with huge wings, swooping down upon headless bodies. [19]

The bird and snake represent the heavens, earth, and Underworld: birds mediate the realms of sky (heaven) and earth, while snakes mediate earth and underworld. Both are potent emblems of regeneration, since birds moult and grow new feathers, while snakes shed their skins and then grow skin anew. Snakes, because of their venom, also represent death. [20] Further, both give birth oviparously: a potent representation of life.

Again, in the early historic age, European and Near Eastern female figures inherited the iconography of the prehistoric bird and snakes. These female figures already had differentiated spheres of influence, although in the earliest historic cultures – the Egyptian and the Sumerian – "Great"-Goddesses were remarkable for their multivalent functionality. Thus the earliest love goddesses were also goddesses of the arts, of fate, and even warrior goddesses.

The Upper Palaeolithic and Neolithic water birds became associated with love goddesses in many early historic cultures. The Sumerian/Akkadian/Babylonian Inanna/Ishtar was depicted with a water bird in a vase from the royal Sumerian city of Larsa. [21] The Greek Aphrodite/Roman Venus was portrayed with swans and geese in many media: she was shown on a clay drinking cup riding a goose, [22] and she was depicted with her young son Eros, riding a swan or goose. [23] The Indic love goddess and goddess of good fortune, Lakshmī, was portrayed with an owl, but an older goddess, the beautiful Rigvedic Sarasvatī, goddess of the river, music, art, and learning, is portrayed riding on a swan (and sometimes on a peacock).

EARLY HISTORIC GODDESSES OF LOVE

Mesopotamian Inanna/Ishtar

In ancient Sumer, the "Great"-Goddess and love goddess *par excellence* was Inanna. She was the morning and evening star. The cuneiform sign for "deity" is a star, and Inanna may at one time have represented the deity *par excellence*.[24] She represented love, war (more so when she became Ishtar among the Akkadians and Babylonians), and most of the prerogatives of society, as represented by the *Me:* she was the

> *Queen of all the me's .*
> *whose hand has gained the seven me's;*
> *My lady: you safeguard the great me's.* [25]

Although this text writes of seven *me's*, it is just used as a magic number. There was a very large number of *me's*, which included the high priesthood, kingship, truth, the dagger and sword, the art of lovemaking, speech, power, and all of the crafts.[26] One of the *me's* was that of judgment, and Inanna decreed the fates of both mortals[27] and gods.[28] She was both beautiful and wise.[29]

Probably Inanna's most important function was that of bestowing love and fertility to humans, animals, and the land. In the exceptionally explicit Sumerian love-poetry her love was described in agricultural and pastoral terminology:

> *... I am the queen of Heaven ...*
> *My husband ...*
> *the wild ox, Dumuzi ...*
> *Inanna ...*
> *[sings] a song*
> *about her vulva ...*
> *'[my] vulva ...*
> *like a horn ...*
> *the ship of Heaven ...*
> *like the new crescent moon*
> *I, the young woman,*
> *who will plant it?*
> *My vulva ...*
> *I, the queen,*
> *who will place the bull?'*
> *'Lady, let the king plant it for [you];*
> *Let Dumuzi, the king,*
> *plant [it] for [you].'"*[30]

In keeping with this agricultural motif, another part of this same fragment describes Dumuzi as

> *. . . the honey of my eye;*
> *he is the lettuce of my heart.* [31]

The descent to the Underworld is related to the ancient agricultural fertility ritual, whereby a goddess, god, hero, or heroine descends to the Underworld in the winter – dies – and is resurrected in the spring, the time when the first blooms, the first crops, begin to grow. Although Inanna is the goddess who descends to the Underworld in the Sumerian myth, nonetheless, when she is saved by the wisdom and water-god Enki, and she rises to the Upper World again, she must designate someone else to be her substitute in the Underworld. When she rises again, she discovers that everyone in the world has been grieving for her, except for her lover, Dumuzi. So she designates him as her substitute. However, Dumuzi's sister, Geshtinanna, asks to spend half the year below, in her brother's stead, and the Inanna decrees that the brother and sister divide up the year, in an alternation of life and death:

> *"Your . . . exactly half the year,*
> *your sister, exactly half the year."* [32]

EGYPTIAN HATHOR

In Egypt, the most multifunctional goddess was Hathor. She represented love and lust: lovers petitioned to her for the favors of their beloveds. She was mistress of the arts: music, crafts, dance, and goddess (and sometimes goddesses) of Fate. She performed an erotic display in order to bring laughter back to her grieving father, the sun-god Rā (here as Prē ʻ-Harakhti):

> *Thereupon, Hathor, Mistress [lit. "lord"] of the Southern*
> *Sycamore tree, went [and] stood before [lit. "in the*
> *foreskin of, in the phallus of"] her father, Master of the*
> *Universe [lit. "Master to the end"], [and] she uncovered*
> *her nether parts before his face, and the great god laughed.* [33]

The god was not laughing *at* his daughter; he was laughing in joy. [34]
Hathor was invoked in Egypt for help in love affairs:

> *I adore the golden lady; I exalt her majesty.*
> *I fashion praise for the Lady of Heaven,*
> *my song of praise for Hathor, the goddess,*
> *I proclaim [my desire] to her.*
> *She gave heed to my prayers.*
> *She directed my mistress to me.* [35]

"Golden" Hathor may have inspired the Greeks to formulate their epithet for Aphrodite. [36] This "golden" color may well relate to the brightly shining planet Venus. Hathor was often represented as a cow or as a woman with the head of a cow, and she may thus have been the inspiration for the epithet "cow-eyed Hera."[37]

INDIC SHRĪ LAKSHMĪ

Shrī Lakshmī [Lakṣmī] was born of the foam at the churning of the ocean,[38] similarly to the Greek Aphrodite's birth in the foam of the sea.[39] The Indic primeval Ocean, the "Ocean of Milk," was an undifferentiated mass of potential energy. The god *Vishnu* [Viśnu], who had metamorphosed into a tortoise, stood in the depths of this ocean; upon his back arose the axis of the world, Mount Mandara, and around Mount Mandara coiled the serpent *Ananta,* the 'boundless one', the 'infinite one'.[40] The creator-god, Brahma, ordered the Hindu gods, or *Devas,* to pull the serpent's head, while the *Asuras,* or demons, alternately pulled his tail. This was a cooperative venture between the two groups of beings, rather than a contest. Pulling in turn upon Ananta's head and tail, the Devas and Asuras caused Mount Mandara to rotate rapidly. In this way they churned the Ocean of Milk. The elements of creation, including Lakshmī and other divinities, came forth out of this Ocean.[41]

Lakshmī is called *Padma,* "goddess of the lotus,"[42] because she was holding a lotus in her

hand when she came up from the ocean; subsequently, the lotus became one of the attributes which she holds in her many hands.[43]

The mythology of Lakshmī was similar to that of Aphrodite in another respect as well: Aphrodite gave birth to *Eros:* literally love-god,[44] and similarly the Indic goddess Lakshmī had a young son, *Kamadeva:* likewise "god of love."[45] Lakshmī was named as consort to various Hindu gods in different texts. In the hymn to the Indic "Great"-Goddess Devī, the *Devīmāhātmyam* — a tale in seven-hundred verses from the *Mārkaṇḍeyapurāṇa*— she was said to be the consort of the god *Shiva* (Śiva):

> *You are indeed Gaurī* (Pārvatī)
> *who has been made the support*
> *of the moon-crested god* (Shiva).[46]

With regard to this assimilation, we recall that the term *devī* means, simply, 'goddess'.

Devī, in her "positive" or more benevolent aspects, was worshipped as *Uma,* the goddess

of "light, splendor."[47]

INDIC DURGĀ

It is possible that the Indic "Great"-Goddess Durgā is connected with Venus, the morning star;[48] this may be the case with the Iranian "Great"-goddess Anahita as well.[49]

HINDU AND BUDDHIST TĀRĀ

The Buddhist Tārā was the compassionate goddess, the powerful savioress and protectress of humanity. She is also found in Hindu texts. Tārā was the wife of Bṛhaspati, the divine priest or sacrificer, the teacher of the gods, who interceded with gods on behalf of humans. Disregarding marital law, the moon-god, Soma, carried off the beautiful Tārā, and this act led to a great war between the gods and the demons. The powerful god, Brahma, finally put an end to the war and restored Tārā to her husband.[50]

Although one possible etymology of the Buddhist Tārā's name is that it derives from Sanskrit *tār*, "cross," *tārayati* "take across, save," it is equally possible that it derives from Proto-Indo-European **ster-*, "star".[51] This would connect her to Akkadian/Babylonian *Ishtar*, West Semitic *Ashtarte*. Monier-Williams' Sanskrit dictionary gives these translations for the adjective *tāra*: "savior, protector, *shining, radiant*, excellent, crossing." As a feminine noun she is "a fixed star."[52]

Tārā has a love aspect: in the fifth book of the *Mahābhārata*, just as the royal seer makes love to Mādhavī (the "honey-woman"), so Bṛhaspati makes love to Tārā.[53]

GREEK APHRODITE

Some classical authors cite the Greek Aphrodite as the goddess of the planet Venus;[54] even though these authors are few, yet her epithet, Urania/Ourania (heavenly), attests to her celestial affiliation. Aphrodite is linked to the heavens through her paternal line: she was born of the foam of the sea when the god Kronos, father of Zeus, castrated his own father, Ouranos. He cast the severed genitals into the sea, and then,

> *A white foam arose from the immortal flesh; in it there grew a maiden.*
> *First she floated to holy Cythera, and from there, afterwards,*
> *she came to sea-girt Cyprus.*[55]
> *There came forth an august and beautiful goddess ...*
> *gods and men call her Aphrodite ...*[56]

Aphrodite is born of the sea just as the planet Venus as evening star appears to rise from the sea into the sky. Further, this watery element (like that of fire) is a source of ever-renewing energy, an energy supplied by the divine feminine. [57]

There are temples of 'heavenly' Aphrodite:

> . . . there is a temple of Aphrodite Ourania. The cult of Ourania was first established by the Assyrians, and after the Assyrians by the Paphians of Cyprus, and the Phoenicians who inhabit Ascalon in Palestine; from the Phoenicians the people of Cythera learned her worship. [58]

Herodotus tells us that

> The temple of Aphrodite Ourania ... [in the Syrian city of Ascalon] is the oldest of all temples erected for this goddess. [59]

We see that the ancient sources believed that Aphrodite came to Greece from the Near East; thus she was borrowed from the Mesopotamian Ishtar or the Syrian Ashtarte. Since Sumerian Inanna/Akkadian-Babylonian Ishtar was called the "Queen of Heaven" (see above), this identification would seem to have been clear from the earliest historic sources of the love goddess.

Aphrodite was *golden,* similarly to the Egyptian Hathor:

> . . .Hermione, who had the appearance of golden Aphrodite. [60]

In *Homeric Hymn* V, Aphrodite represented the "Great-Mother," Cybele. On her journey to her liaison with the mortal, Anchises (father of Aeneas), she was accompanied by

> gray wolves, wagging their tails,
> and bright-eyed lions, and bears,
> and swift panthers, greedy for roe-deer; and she, seeing them,
> delighted in her soul,
> and she put desire into their hearts,
> so that they all lay down, two by two, in shady dens. [61]

The love goddesses — Inanna/Ishtar, Syrian Anat, Greek Aphrodite — have another myth in common. They all love their beloveds, and they grieve when their lovers die: in the Descent of Inanna, Inanna curses Dumuzi to inhabit the Underworld, but in other poems she laments for him. [62] Anat grieves for Ba'al after the latter is killed by the god of the Underworld, Mot;[63] Isis, griefstricken, searches the world for her husband Osiris, after he has been killed by the Underworld god Seth.

The Greeks borrowed much myth from their Near Eastern neighbors, and the myth of Demeter searching for her daughter Persephone probably owes its core to the Egyptian myth of Isis. Persephone too must die (that is, spend the winter season of the year in the Underworld) and live again (when she is in the Upper World).[64] But the Greeks also borrowed a myth of the love-goddess grieving over her dead lover. Aphrodite laments the handsome mortal, Adonis (Syrian rather than Greek in origin), after he has been killed by a wild boar in a hunting accident.[65] However, Adonis in origin is no mortal; his name is related to the Semitic word for 'Lord', as found in the Hebrew Old Testament.

In the Troad, Aphrodite Aeneas may have been a form of the Phrygian "Great"-Goddess, Cybele.[66] In some myths of Cybele, Cybele's priest, Attis, is conflated with the Syrian/Greek Adonis. In most versions of the myth, Attis dies after emasculating himself. But in the Lydian version,[67] dating to the 6th century BCE, Attis was killed by a boar, just as Adonis; here Attis was syncretized to the dying and resurrecting consort-god. However, the 'mystery' religion of a resurrecting Attis is quite late.[68] In the later Empire, there are instituted festivals in his honor: the Tristia, sad rites commemorating his death, and the Hilaria, joyous feasts in honor of his resurrection.

The Greek bucolic poet Bion (ca. 100 BCE) tells us that

> *Aphrodite, having unbound her hair, [wanders], wailing in the woods, mourning, with unbraided hair, barefoot...calling for her Assyrian husband...*
> *[Refrain:] "Alas, Adonis."*[69]

It is probably no accident that the love-goddess, the goddess of the planet Venus, wanders in her myths. Etymologically, a planet is a wanderer (from Greek ἀστήρες πλανήται, astēres planētai, "stars wandering" > "wandering stars"). So just as the planet Venus is etymologically a 'wandering star', thus it is appropriate that the love-goddess, the Venus-goddess, wanders – in this case, searching for her lover.

The myths vary, but the goddess searches for her lover — wanders — and grieves.[70] There is a fragment of a poem which alludes to a story about the Aeolian poet Sappho jumping into the sea because of her love of the handsome youth, Phaon.[71] Sappho was a female poet who was exceptionally well-regarded by both the ancient Greeks and Romans. It is inferred that Sappho jumped off the White Rock of Leukas. There is a parallel between the youths Adonis and Phaon, and parallelism between Sappho leaping off a rock into the ocean for love of Phaon and Aphrodite who, as evening star, dives after the Sun after it has set.[72]

Aphrodite on Lesbos

About a decade ago I brought my daughter, Leah, to a wonderful workshop on the island of Lesbos, home of Sappho. This was an annual workshop conducted by the Ariadne Institute and its founder, the scholar Carol P. Christ. A night or so after the workshop, we were relaxing on the outside deck of Carol's home in Molivos, and Carol remarked on the rising of the evening star, Venus (Aphrodite). My daughter's response, which I think is magical, was "Oh! She is everywhere!"

ROMAN VENUS

The planet Venus and the spring-month of April were likewise sacred to the goddess Venus. (Ovid, *Fasti* iv. 89-90.)

The Romans also borrowed the Egyptian goddess Isis, who in the later era syncretized several Egyptian goddesses. According to the Roman author Apuleius (who was born in Algeria),

> *Oh, queen of heaven, whether you be the nurturing Ceres or . . . the celestial Venus .. . or . . . the sister of Phoebus [Diana] or . . . Proserpine with triple face . .by whatever name or rite or shape it is lawful to call upon you.*[73]

Venus in origin was probably an old Italic goddess representing love. Her Indo-European name, *Venus,* is related to the Sanskrit term for "loveliness" *vanas,* the Old Icelandic word for "friend" *vinr,* feminine *vina;* and the Hittite verb *wenzi,* "s(h)e has sexual intercourse".

SCANDINAVIAN FREYJA

Freyja, "the lady," was connected with both war and with love, similarly to other "Great"-Goddesses from many other cultures. We are told that whenever she rides into battle, to her lot falls half of the slain warriors.[74] Her chariot is drawn by two cats,[75] animals which may represent the earlier Neolithic European lions which accompanied some of the goddesses.[76]

Freyja had a falcon coat, and she often lent her 'feather form' to others.[77] Thus, as a goddess who could take on bird-shape, she was linked to the Neolithic bird-goddess.

Her most important function seems to be that of love-goddess. Love-poetry is pleasing to her; she is invoked for help in love-affairs.[78] Her husband Oðr (perhaps a form of Oðin) goes on long journeys, and when he is gone she weeps tears of red-gold;[79] she searches for him among peoples she does not know. Thus she wandered and grieved for her missing lover, similarly to Inanna, Ishtar, Aphrodite, and others.

In keeping with her function of fertility and nurturing, Freyja is sometimes called *Gefn,* "giver" and *Syr,* "sow."[80] She was thus the mother *par excellence,* as was the sow.

Because the Scandinavian Freyja was goddess of love, the satirical poem, the "Lokasenna," accuses her of promiscuity:

> *Freyja ...all Aesir and alfs who are here within*

each one have you embraced, you adulteress![81]

This poem was a highly satirical piece, and we may assume that the poet simply desires to cast aspersion upon the goddess. However, in order to obtain her famous golden necklace from the four dwarves who crafted it, she was required to pass one night with each of them, and she did so.[82] Thus, it is clear that Freyja was connected with both love and sexuality.

Although there are no texts which relate specifically that Freyja was associated with the planet Venus, one of the mountains of Venus is called Freyja Montes.[83] Further, the Germanic name for the day of the week, *Friday*, "day of Freyja," is related to the Roman day of the week, *dies Veneris*, "day of Venus."

CONCLUSIONS

It would appear that one of the first functions of the divine feminine (in particular, the divine feminine genitalia) was apotropaic: that of guarding and protecting a space. By the time of the earliest texts, the "Great"-Goddesses — goddesses whose functions were manifold: love, war, and often many other functions as well — are also the goddesses of the morning and evening star: the planet Venus. They wander — as a planet does — in search of their loves. The association of the love-goddess with the planet Venus appears particularly in the Near East, in the earliest historic cultures of Egypt and Mesopotamia. The Greeks borrowed Aphrodite from the Near East — probably from the Western Mediterranean Ashtarte, borrowing the celestial function of the goddess as well. The Romans, in borrowing Greek deities, kept the astral function for Venus, originally probably a love-goddess without astral function, as well. The Indic Durgā and Tārā too have astral functions. Thus a celestial function is associated with love-goddesses in the very earliest historical texts.

WORKS CITED

Apuleius (Rudolf Helm, ed.). *Apulei Opera Quae Supersunt.* Series: Bibliotheca scriptorium graecorum et romanorum Teubneriana. Leipzig: Teubner. 1908-1913.

Aristophanes. Rogers, Benjamin Bickley, ed. *Aristophanes.* Cambridge: Harvard University Press. Vol III - Lysistrata. 1972.

Bion. Edmonds, J.M., ed. *The Greek Bucolic Poets.* Cambridge: Harvard University Press. 1977. Chantraine, Pierre. *Dictionnaire Étymologique de la Langue Grecque: Histoire des Mots.* Paris: Klincksieck. 1968-1980.

Conard, Nicholas J. "A female figurine from the basal Aurignacian of Hohle Fels Cave in southwestern Germany," *Nature* 459 [14 May 2009]: 248–252. 2009.

Dexter, Miriam Robbins. *Whence the Goddesses: A Source Book.* New York: Pergamon. Athene Series. (Teachers College, Athene Series) 1990.

———. 1997a. "The Frightful Goddess: Birds, Snakes and Witches" in *Varia on the Indo-European Past: Papers in Memory of Marija Gimbutas*: 124-154. Miriam Robbins Dexter and Edgar C. Polomé, eds. (Journal of Indo-European Studies Monograph #19). Washington, DC, The Institute for the Study of Man, 1997.

———. 1997b. "Born of the Foam" in *Studies in Honor of Jaan Puhvel, Part II: Mythology and Religion:* 83-102. John Greppin and Edgar C. Polomé, eds. *Journal of Indo-European Studies* Monograph 21. Washington, DC: Institute for the Study of Man.

———. 2009. "Ancient Felines and the Great-Goddess in Anatolia: Kubaba and Cybele." In *Proceedings of the 20th Annual UCLA Indo-European Conference*. Stephanie W. Jamison, H. Craig Melchert, and Brent Vine (eds.). 2009, 53–67. Bremen: Hempen.

———. Forthcoming. "Substrate continuity in Indo-European Religion and Iconography: Seals and Figurines of the Indus Valley Culture and Historic Indic Female Figures." *Journal of Indo-European Studies* Monograph.

Dexter, Miriam Robbins and Victor H. Mair. *Sacred Display: Divine and Magical Female Figures of Eurasia*. Amherst, New York: Cambria Press. 2010.

Eddas, Poetic. Kuhn, Hans. *Edda: Die Lieder des Codex Regius Nebst Verwandten Denkmälern* I-II. Heidelberg: Carl Winter. 1962-1968.

Eddas, Prose. Faulkes, Anthony, ed. *Snorri Sturluson, Edda, Prologue and Gylfaginning*. Oxford: Clarendon Press. 1982.

Farnell, Lewis R. *Cults of the Greek States*. Vol. II. Oxford: Clarendon Press. 1896.

Frisk, Hjalmar. *Griechisches Etymologisches Wörterbuch* I-II. Heidelberg: Carl Winter. 1960.

Gardiner, Sir Alan H. *The Library of A. Chester Beatty: The Chester Beatty Papyri* I. London: Walker (Oxford University Press). 1931.

Gimbutas, Marija. *The Language of the Goddess*. San Francisco: HarperSanFrancisco. 1989.

———. 1991. *The Civilization of the Goddess. The World of Old Europe*, edited by Joan Marler. San Francisco: HarperSanFrancisco.

———. 1999. *The Living Goddesses*, by Marija Gimbutas, edited and supplemented by Miriam Robbins Dexter. Berkeley/Los Angeles: University of California Press.

Haarmann, Harald and Joan Marler. *Introducing the Mythological Crescent: Ancient Beliefs and Imagery Connecting Eurasia with Anatolia*. Wiesbaden: Harrassowitz Verlag. 2008.

Hallo, William W. and Van Dijk, J.J.A., eds. *Exaltation of Inanna*. New Haven: Yale University Press. 1968.

Harivaśa. Vaidya, Parasharam Lakshman, ed. *Harivaśa*. Poona: Bhandarkar. 1969-71.

Jacobsen, Thorkild and Kramer, Samuel N. "The Myth of Inanna and Bilulu." *Journal of Near Eastern Studies* XII.3: 160-188. 1953.

Kramer, Samuel Noah. "Cuneiform Studies and the History of Literature: The Sumerian Sacred Marriage Texts." *Proceedings of the American Philosophical Society* (*PAPS*) 107.6: 485-516. 1963.

———. 1979. *From the Poetry of Sumer*. Berkeley: University of California Press.

Le Martin, Max. "On the Footsteps of the Lajja Gauri in the ancient kingdoms of the Near East." http://www.scribd.com/max8899/d/55983753-On-the-Footsteps-of-Lajja-Gauri-in-the-Ancient-Kingdoms-of-the-Near-East-Completed. 2010.

———. 2010. "Lajja Gauri-like in the Indus Valley." http://www.scribd.com/doc/33492538/30970713-Lajja-Gauri-like-the-erotic-goddess-of-the-Indus-valley-Completed.

Liddell, Henry G. and Scott, Robert. *Greek-English Lexicon*. 9th Edition. New York: Harper and Brothers. 1856 [1961].

Los Angeles Times. June 25, 2009.

Lucian. *De Dea Syria*. ("About the Syrian Goddess," attributed to Lucian), Harold W. Attridge and Robert A. Oden, eds. (English and Greek) Missoula, Montana: Scholars Press for the Society of Biblical Literature. 1976.

Mahābhārata. Sukthankar, Vishnu S., ed. *The Mahābhārata*. Poona: Bhandarkar. 1940.

Mahābhārata. Van Buitenen, J.A.B., ed. and transl. *The Mahābhārata*. Chicago and London: The University of Chicago Press. 1978.

Monier-Williams, Sir M. *Sanskrit-English Dictionary*. Oxford: Clarendon Press. 1899.

Nagy, Gregory. "The White Rock of Leukas." *Harvard Studies in Classical Philology* 77 (1973): 175.

Ovid (Naso, Publius Ovidius) (Merkel, R, ed.). *Ovid, Works*. Leipzig: Teubner. 1907.

Parpola, Asko. *Deciphering the Indus Script*. Cambridge: University Press. 1994.

Pritchard, James B., ed. *Ancient Near Eastern Texts Relating to the Old Testament (ANET)*. Princeton, New Jersey: Princeton University Press, Third Edition. 1969.

Rāmāyaṇa. Mudholakara, Shastri Shrinivasa Katti, ed. *Rāmāyaṇa of Vālmīki*. New Delhi: Parimal. 1983

Sappho. Page, Denys, ed. *Sappho and Alcaeus*. Oxford: Clarendon Press. 1955.

Vermaseren, Maarten J. transl. from the Dutch by M.H. Lemmers. *Cybele and Attis: the Myth and the Cult*. London: Thames and Hudson. 1977.

Saundaryalaharī. Brown, W. Norman, ed. *The Saundaryalaharī, or Flood of Beauty*. Cambridge: Harvard University Press. 1958.

Westenholz, Joan and Aage Westenholz. "Help for Rejected Suitors: The Old Akkadian Love Incantation MAD V 8." *Orientalia* 46: 198-219. 1977.

Wolkstein, Diane and Kramer, Samuel Noah (1983): *Inanna: Queen of Heaven and Earth*. New York: Harper and Row.

http://en.wikipedia.org/wiki/List_of_montes_on_Venus.

NOTES

1. See Conard: 2009.
2. See the *Los Angeles Times*, June 25, 2009.
3. Bradshaw Report: http://www.bradshawfoundation.com/chauvet/venus_sorcerer.php, on the Venus and The Sorcerer (the bison): "The Venus is the earliest of the designs. The feline on the left, the Sorcerer, and the multiple lines on the right, are all painted or engraved later."
4. Bradshaw Report: http://www.bradshawfoundation.com/chauvet/venus_sorcerer.php: "Perhaps the female representation relates directly to the corridor to the chamber, which opens just behind her. Four other female representations limited to just the pubic triangle are in the cave; they are all in the system including the Galerie des Megaceros and the Salle du Fond, indicating each time the entrance to the adjacent cavities...A cluster of convergent data suggests that the Venus is Aurignacian and that she was created in the first period of the decoration of the Chauvet Cave."
5. See Dexter 2009.
6. I thank Joan Marler for this insight: "She is the earth beneath our feet, and her body is the temple." Personal Communication, September 2, 2011.
7. See Dexter and Mair 2010.
8. Haarmann and Marler 2008: 24, Figure 4: Upper Palaeolithic Cave Painting: Roof Stone from Peri Nos, Lake Onega, Republic of Karelia.
9. Max Le Martin. "In the Footsteps of the Lajja Gauri." Online article.
10. See Haarmann and Marler 2008, Figure 16: female shaman in magical dance, with bow; in a rock painting from Astuvansalmi (Ristiina, Finland).
11. Garfinkel 2003.
12. Ankara. Anatolian Civilizations Museum. Ana Tanrica ("Mother Goddess") excavated from Çatalhöyük.
13. National Archaeological Museum, Greece. N.M. 3928.
14. For hundreds of examples of Neolithic female figures, including images, see Gimbutas 1989, 1991, 1999.
15. http://www.tantraworks.com/Ancient_Tantra.html#indus. March 14, 2006. I thank Vicki Noble for this reference.
16. Le Martin, n.d., "Lajjā-Gaurī-like".
17. See Dexter Forthcoming.
18. See Dexter 1997a.
19. Mellaart (1965): 98; 101, figure 86. These representations were found on Level VII of the site.
20. See Dexter 1997a.
21. Painted Terracotta. Ht. 10 3/8". 1900-1800 BCE. Louvre AO17000.
22. Clay drinking cup (*kylex*) from Rhodes, Attic, *ca.* 460 BCE. British Museum D2; perhaps a prototype for the nursery character, *Mother Goose*.

23. Terracotta figurine, Tarentum, Italy; *ca.* 380 BCE, British Mus. 1308
24. Note that in the South Indian Tamil language, the term *veḷḷi* is used for both Venus and 'star' in general. See Parpola 1994: 231. Again, Venus may be the star (planet) which represents the whole of the firmament.
25. All translations in this chapter are by the author. Enheduanna, "Exaltation of Inanna" 1; 5-6; *ca.* 2500 BCE. Hallo, ed. (1968) lines 1; 5-6:
nin-me-šár-ra . . .
me-imin-bé
*šu sá-du*ll-*ga*
nin-mu me-gal-gal-la
sag-kešda-bi za-e-me-en.
26. See Wolkstein and Kramer (1983): 16-18.
27. See Kramer (1979): 86.
28. See Kramer (1979): 89-90.
29. See Dexter (1990): 21; 197, notes 59-60.
30. Sumerian Fragment; Recorded 2000-1500 BCE. Text is in Kramer, (1963): 505, lines 5; 8-9; 16-21; 26-30:
. . . ga-ša-an-an-na me-[e] . . .
mu-ud-na-mu . . .
am- ddumu-zi . . .
d *inanna-ke4 . . .*
SAL-la-ni sir-ra . . .
SAL-la . . .
si-gim . . .
ma-an-na .
u4 sar-gibil-gim . . .
ki-sikil-mèn a-ba-a urx-ru-a-bi
SAL-la-mu . . .
ga-ša-an-mèn gu4 a-ba-a bi-íb-gub-bé
in-nin9 lugal-e a-ra-an-urx-ru
d*dumu-zi lugal-e a-ra-urx-ru.*
31. See Kramer, (1963): UET VI, No. 10: 508, line 21:
*igi-mà làl-bi-im šà-mà i-(is)*sar-*bi-im.*
32. Sumerian Fragment; Recorded 2000-1500 BCE. Text is in Kramer 1963: 515, line 10:
. . . -za mu-maš-àm nin9-*zu mu-maš-àm.*
33. *Papyrus Chester Beatty* I, Recto, 4.2–4.3. The Egyptian hieroglyphic text is in Gardiner, 1931. Determinatives have not been included in the following transliteration:
Het..her.r.t neb n.h.a res .er i.a.ii i.w.s.t. .er āha.ā m met i t.f.s.t
neb.r dr i.w.s.t kef.a.w.t.ph.t. ka.t.s.t.r .er.f a.a.ā.n neter āa
seb.i.a.i.m.s.t.
34. For other female figures which enact the "sacred display" or *anasyrma* (the "lifting up of the skirts) see Dexter and Mair 2010.
35. *Papyrus Chester Beatty* I, verso C, page 3 (plate XXIV). New Kingdom, 1550–1080 BCE. For the Egyptian text, see Gardiner (1931):
Nubit s-šua-a· .em set s-qai-a·
nebt pet a·ri-a· a·ut
en .et-.ert .eknu-a· en .enut
sma· -a· en-s
setchem set speru-a·
utch set en-a· .enut.
36. For example, see Homer, *Odyssey* 4.14 and see below.
37. See for example Homer, *Iliad* 1.551.
38. *Rāmāyaṇa* 1.45. *Mahābhārata* 1.16.34. *Pañcarātra* III. There are other stories of her birth: cf. *Śatapatha Brāhmaṇa* III.1.
39. See Dexter 1990 ix.
40. See Monier-Williams (1899): 25.

41. *Rāmāyaṇa* I.45.
42. *Mahābhārata* I.92.26-27. *Saundaryalaharī* 98.
43. 'Meditation of Mahālakṣmī', preceding *Devīmāhātmyam* I.
44. Cf. Sappho, Fragment 198 [Scholia on Theocritus xiii.1-2c, in Page (1955): 105]:
Ἔρωτα...Ἀφροδίτης καὶ Οὐρανοῦ
"Eros, son of Aphrodite and Ouranos."
Early on, Eros was given widely differing genealogies; even Sappho gave him two different sets of parents. Later, Eros, and particularly his Roman counterpart, Cupid, became more widely accepted as son of Aphrodite/Venus. See Ovid, *Metamorphoses* 1.463; IX.482; Apuleius, *Metamorphoses*, IV.30-31 *et passim;* Horace, *Carmina* 1.19.1. Eros had both a physical and a *cosmic* significance; in the latter, he was one of the three primeval beings, along with Chaos and Earth, which were created before all else. Cf. Hesiod, *Theogony* 120.
Linguistically the Greek word *eros* does not have a well-accepted etymology. See Chantraine (1968-1980: 364), under ἔραμαι. See also Frisk (1960: 547). Thus the word, as well as the deity, was probably a borrowing from another ancient language. A good candidate is Akkadian: in an Old Akkadian love charm, dating to the third millennium BCE, the goddess Ishtar (in Sumerian, Inanna) is the mother of the love-god Er'emum:
...Er'emum, the son of Ishtar, sitting in her lap....
The Old Akkadian text, dating to ca. 2200 BCE, is in Gelb (1970), *Materials for the Assyrian Dictionary* (MAD) 5: 8.3-4:
ir-e-mu-um *DUMU* dInanna in za-gi-[sa uša]b
The translation "sitting in her lap" is from Westenholz and Westenholz (1977: 203). Old Akkadian *Ir'emun/ Er'emun*, the name for the love-god, is built upon the root *ra'amu*, 'to love'. See Dexter 1997b: 96. Westenholz and Westenholz (1977, 205–207) cite several instances of *ir-ri-mu* in the Old Babylonian literary texts; this they hold to be the same as Er'emum. So the god probably has a long, continuous history.
45. Harivaṁśa II. App. 1.41.525: 541:
dharmāllakṣmyudbhavaḥ kāmaḥ
46. *Devīmāhātmyam* IV.11; *ca.* 550 CE.
gaurī tvameva śaśimaulikṛtapratiṭā.
47. Monier-Williams: 217; *Mahābhārata* I.207.18 *et passim;* *Saundaryalaharī* 71; App. 2.
48. See Parpola 1992: 231.
49. Parpola 1994: 262; 231.
50. *Mahābhārata* V.3972; *Harivaṁśa* 1340 ff.
51. Parpola 1994: 262 cites Subandhu's Vāsavadattā (ca. 650 BCE): "The Lady Twilight was seen, devoted to the stars and clad in red sky, as a Buddhist nun [is devoted to Tārā and is clad in red garments]."
52. Monier-Williams 443, col. 3.
53. *Mahābhārata* V.3972 (van Buitenen edition 115.15).
54. See Liddell and Scott (1961: 293, col. 2), citing Plato, "Epinomis" 987b.
55. Cythera lies off the Laconian coast of the southeastern Peloponnese; Cyprus is a large island located south of Cythera.
56. Hesiod, *Theogony* 190-197; ca. 750? BCE:
...ἀμφὶ δὲ λευκὸς ἀφρὸς ἀπ᾽ ἀθανάτου χροὸς ὤρνυτο·
τῷ δ᾽ ἔνι κούρη ἐθρέφθη· πρῶτον δὲ Κυθήροισιν ζαθέοισιν ἔπλητ᾽,
ἔνθεν ἔπειτα περίρρυτον ἵκετο Κύπρον.
ἐκ δ᾽ ἔβη αἰδοίη καλὴ θεός...τὴν δ᾽ Ἀφροδίτην
κικλήσκουσι θεοί τε καὶ ἀνέρες...
57. Many ancient female figures are associated with water, for example, as rivers such as the Brighid in Ireland or the Seine in Paris (from the Roman goddess Sequana). Perhaps the element of water, ever-flowing, is related to a notion of perpetual – divine – energy. The goddess is connected with perpetual fire as well – and the particularly bright Venus planet would appear to be a perpetual fire. See Dexter 1990: 165.
58. Pausanias, *Description of Greece* 1.14.7; *ca.* 150 CE:
...ἱερόν ἐστιν Ἀφροδίτης Οὐρανίας...πρώτοις δὲ ἀνθρώπων Ἀσσυρίοις
κατέστη σέβεσθαι τὴν Οὐρανίαν μετὰ δὲ Ἀσσυρίους Κυπρίων Παφίοις

καὶ Φοινίκων τοῖς 'Ασκάλωνα ἔχουσιν ἐν τῇ Παλαιστίνη,
παρὰ δὲ Φοινίκων Κυθήριοι μαθόντες σέβουσιν·
   59. Herodotus, *The History* 1.105; *ca.* 450 BCE:
...τῆς οὐρανίης 'Αφροδίτης τὸ ἱρόν...πάντων ἀρχαιότατον ἱρῶν
ὅσα ταύτης τῆς θεοῦ·
   60. Homer, *Odyssey* 4.14; ca. 800? BCE
Ἑρμιόνην, ἣ εἶδος ἔχε χρυσέης 'Αφροδίτης.
   61. *Homeric Hymn* V, to Aphrodite: 70-74; 8th to 6th Centuries BCE.
σαίνοντες πολιοί τε λύκοι χαροποί τε λέοντες,
ἄρκτοι παρδαλιές τε θοαὶ προκάδων ἀκόρητοι
ἣ δ' ὁρόωσα μετὰ φρεσὶ τέρπετο θυμὸν
καὶ τοῖς ἐν στήθεσσι βάλ.' ἵμερον· οἳ δ' ἅμα πάντες
σύνδυο κοιμήσαντο κατὰ σκιόεντας ἐναύλους.
   62. For example, see the myth of "Inanna and Bilulu," in Jacobsen and Kramer (1953); the authors assign the text to the Bad-tibira branch of the Dumuzi tradition. One of the main motifs of the text, the death of Dumuzi in a raid on his sheepfold, is found in Badtibiran myth. In Jacobsen's version of this myth, the death of Dumuzi is caused by an old woman, Bilulu. Inanna goes to visit Dumuzi in his sheepfold one day and is informed that he is dead. Inanna fashions a song of both praise and lament for him, and then she wreaks vengeance upon the woman by killing her and turning her into a waterskin, so that refreshing cold water will sweeten the resting place of Dumuzi. (Samuel Noah Kramer [1953: 187-188] believes that it was not Bilulu who killed Dumuzi in this myth.) According to the text, after Inanna discovers that Dumuzi is dead she composes a song of both praise (1953: 175, lines 80 ff) and lamentation (1953: 179, lines 165 ff) for him. The significant statement of this article is that the text was composed for use at the yearly ritual of lamentation of Dumuzi in that city. (1953: 163) There were several Sumerian deities who, like Dumuzi, were alive for only part of the year (1953: 186), which necessitated a seasonal death and then resurrection. For discussion of the myth see also Kramer (1979): 83-84.
The Akkadian/Babylonian Ishtar was known as the "star of lamentation": see Ferris J. Stephens, "Prayer of Lamentation to Ishtar," in J. Pritchard, ed. (384, line 9.) She is also the "light of heaven and earth" (line 5), the "lady of heaven and earth," (line 27), and the "torch of heaven and earth." (Line 35)
   63. See Wolkstein and Kramer (1983): 16-18.
   64. See Dexter 1990: 126-127.
   65. See Apollodorus, *Atheniensis Bibliothecae*, III.xiv.4. The *Adonia* or feasts of Adonis were celebrated in the Classical era. His rituals were composed of ceremonial mourning and the singing of dirges. See also Lucian (2nd century CE): 6 on lamentations for Adonis in the Phoenician city of Byblos (modern day Lebanon). Aristophanes, *Lysistrata* 393, writes, "the woman, dancing [on the roof], cried, 'alas, Adonis!' ἡ γυνὴ δ' ὀρχουμένη 'αἰαῖ Ἄδωνιν' φησίν. An early testimony to the Adonia, in Lesbos, is provided by fragmentary lines of Sappho: "graceful Adonis is dying, Oh Kytherea [that is, Aphrodite]" and she tells the young girls to beat their breasts. (D 107, E 103, LP 140: κατθνα[ί]σκει, Κυθέρη', ἄβρος Ἄδωνις·)
   66. See Farnell 1896 II.641. Farnell also believed that the Cretan Aphrodite was a divinity closely akin to Cybele (II.633).
   67. Herodotus I. 34-45. In this account, Attis was the son of the wealthy Croesus, the last king of Lydia. Croesus had a dream, wherein Attis was killed by a spear of iron. Trying to protect his son, he caused him to be married. Afterwards, Attis prevailed upon him to allow him to take part in a boar hunt; Attis was killed when the Phrygian Adrastus, whom Croesus had taken into his house, threw a spear meant for a boar and, missing the boar, instead hit Attis.
   68. Vermaseren 1977: 122-23.
   69. Bion (ca. 100 BCE) I.19-32:
ἁ δ'Ἀφροδίτα
λυσαμένα πλοκαμῖδας ἀνὰ δρυμὼς ἀλάληται
πενθαλέα νήπλεκτος ἀσάνδαλος·
... 'Ασσύριον βοόωσα πόσιν...
... 'αἲ τὸν Ἄδωνιν.'

70. See Dexter (1990): 21.
71. See Nagy 1973: 141.
72. For discussion, see Nagy 1973: 142-143. Aphrodite is here the planet 'Venus: "By diving from the White Rock, she [Sappho] does what Aphrodite does in the form of Evening Star, diving after the sunken Sun in order to retrieve him the next morning in the form of Morning Star." (175) Sappho projects her identity into the goddess. Farnell 1896 II.637 finds ceremonies of mourning in the worship of the sea-nymph Leucothea at Thebes, and he connects Leucothea and Aphrodite.
73. Apuleius, *Metamorphoses* XI.2; born *ca.* 123 CE:
Regina caeli, sive tu Ceres alma .. .seu tu caelestis Venus .. .seu Phoebi soror .. .
seu . . . Proserpina triformi facie .. .quoque nomine, quoque ritu, quaque facie,
te fas est invocare ...
74. Poetic Edda, "Lokasenna" 30. See also the Poetic Edda, "Grimnismal" 14; Snorri, "Gylfaginning" 24.
75. Snorri, "Gylfaginning" 24.
76. See Dexter 2009.
77. In Poetic Edda, "Þrymskviða" 3, Thor (Þor) asks Freyja to lend him her *fiaðrhams,* her 'feather garment'.
78. Snorri, "Gylfaginning" 24.
79. Snorri, "Gylfaginning" 36.
80. Snorri, "Gylfaginning" 36.
81. Poetic Edda, "Lokasenna" 30; *ca.* 1000 CE.
. . . Freyia! .. .
ása oc álfa er hér inni ero, hverr hefir þinn hór verið.
82. "Sørla Þattr" 1.
83. http://en.wikipedia.org/wiki/List_of_montes_on_Venus.

Chapter Four

Venus, The Caillichín na Mochóirighe of Newgrange, Ireland

Anthony Murphy

On the east coast of Ireland, not far from the sea, lie the remnants of a vast monumental landscape, laid down by a community of astronomers and farmers who lived along the shores of the Boyne River over 5,000 years ago.

These remains, concentrated in a bend in the Boyne a few miles upstream of the town of Drogheda, represent the zenith of achievement of a remarkable and ancient people, who enshrined their science and spiritual beliefs immortally in stone.

The largest of the Boyne Valley monuments are Newgrange, Dowth and Knowth, each seemingly grander and more lavishly constructed than their counterparts, and each steeped in myth and cosmology.

This is a very ancient landscape, grafted by ice sheets over a period of a million and a half years, ending in the last great Ice Age 10,000 years ago. When our distant ancestors came here to clear the forests from the banks of the Boyne, they had already mastered significant skills which would aid them in their great work, to construct these giant edifices which would immortalise their efforts.

They were farmers, who had learned to sow crops, and to keep cattle. Their agrarian skills enabled them to progress from the "hunter gatherer" lifestyle which had kept their forefathers so occupied so much of the time as to disallow any further pursuits. Farming the land meant there was a sustainable food supply, and that not everything had to be caught in the forests and on the hills.

The looping bend of the Boyne River forms something akin to a giant island of land upon which the great monuments of Newgrange, Knowth and Dowth sit, Newgrange the most prominent of the three. The area has some of the richest farmland in Ireland, its fertile soil continuing to produce lush and plentiful crops as it has done for thousands of years.

Here, in what archaeologists call the Neolithic, or New Stone Age, these early farmers devoted themselves to the grandest of projects, the construction of giant mounds of earth and stone. We know from examination of these remarkable monuments that the people who inhabited the Boyne Valley in prehistory were not just farmers. They had developed a significant set of skills, which included surveying, engineering and astronomy.

Little of their capabilities was really understood, at least in academic circles, even long after the first tentative steps were taken towards exploring these giant structures. It was not until the latter part of the 20th century that we started to build a greater picture of these sites, and the people who put them together. The mounds were not mere tombs, built solely for the purpose of burying the bones, burnt and unburnt, of the dead. They had a much greater purpose than this, in fact a multiplicity of functions and intentions which it is not easy to describe with one term. Professor Michael O'Kelly, who began excavating Newgrange in the 1960s, rediscovered the Winter Solstice phenomenon, during which light from the rising sun on the shortest days penetrates the long stone corridor of Newgrange and illuminates the central chamber deep inside. This marked the beginning of the modern inter-pretation of the Stone Age passage-mounds, not only those of the Boyne, but passage-mounds in other areas of Ireland, in particular Sligo and Loughcrew. As excavations progressed at both Newgrange, and nearby Knowth, a stun-ning overview of a wonderful and talented people emerged.

The builders of the Boyne monuments were not Neanderthals, not the rude pagans the British antiquarians would have had us believe. They were a focused and dedicated people, who lived close to the land and nature, and whose cosmic vision unfolded during the course of their monumental con-struction pursuits. They had built three giant passage-mounds, each contain-ing hundreds of thousands of tonnes of stones and earth, and constructed them in such a way as to ensure they would endure the ravages of time and nature to survive long into the distant future. The water-proofing mecha-nisms put in place at Newgrange, for instance, kept the interior largely dry until recent times, except for one leak in the eastern recess of the chamber.

The solstice event at Newgrange was heralded internationally as a mo-mentous discovery — and rightly so — but the local people who lived in the Boyne Valley in recent history, people who might well have been direct descendants of the builders, never really forgot what Newgrange was about. It had long been recounted in various stories that Newgrange was aligned so that the sun shone into its chamber on the solstice. This was known long before Professor O'Kelly first put trowel to sod in his efforts to reveal the secrets of Newgrange to the world, and long before he rediscovered the event in the winter of 1967.

Therein lies one significant key to the unlocking of the secrets of the stones — the myths, legends and folklore of Ireland. The people of this island have a very ancient memory, knowledge of events and happenings places and people from long, long ago. For most of our history as a people, we have preserved knowledge, some of it scientific, in the form of tale and verse, song and poem. The survival of knowledge, so often ascribed only to the Christian monks who preserved our ancient stories by transcribing them onto parchment in the 12th and 13th centuries, is due in greater part to the endurance of myth. Our most ancient memories as a people are locked into these great tales, which we have been uttering for countless generations, wittingly and unwittingly bringing ancient stories into modern times. Examination of the stories of the Boyne Valley helps greatly in the unravelling of a complex mystery. Those who rely on scientific data and ancient artefacts without considering the myths will undoubtedly miss part of the picture. It is only where stones, stars and stories come together that we can begin to better understand our distant forebears and their ancient endeavours.

Collectively, the monumental landscape is known in Irish mythology as Brú na Bóinne, and in this name the supreme goddess of the ancient valley reveals herself. She is Bóann, or Bóinn, meaning White Cow, from *bó* (cow) and *finne* (Bright/White).

Bóinn is said to be the wife of Nechtain, another principal deity of the Tuatha Dé Danann, the ancient gods. She gives her name to the river Boyne, which is Anglicised from her Irish name.

We know from osteoarchaeology, the study of ancient bones, that cattle were kept by the builders of Newgrange and the great mounds. In fact it is the practice of grazing cattle combined with the growing of crops and the production of pottery that archaeologists consider the traits that mark out the Neolithic or New Stone Age. Previously, in the Mesolithic, people relied on hunting, fishing and foraging for food and were not engaged in agrarian practices. The agricultural developments that had taken place at the beginning of the new age, the Neolithic, are a big factor in the construction of the monuments. For the first time, people did not have to spend all their time chasing the game. The production of cereals and the availability of milk and meat from cattle meant that more time could be spent on other pursuits, including the construction of the monuments.

The importance of cattle to the monument builders cannot be overstated. At some early stage it was recognised that the bovine gestation period is the same as that of humans. And it is highly likely, given that the Neolithic builders were accomplished astronomers and their ancestors had long recognised the patterns and cycles of the heavenly bodies, that it was recognised in far-off times that the bovine gestation was equal to nine and a half synodic lunar months.

There is a plethora of bovine mythology in the ancient Irish texts. Much of it can be found in the Boyne Valley, but stories about magical cows abound in different parts of Ireland. One tale that crops up on both west and east coasts features a magic cow and her calf which are stolen by the great Irish "Cyclops," Balor. Realising they are being kidnapped, the cow screams and Balor opens his giant, single eye, causing the cow and calf to be petrified into rocks. The story is found on the east coast where the Boyne River enters the sea, some 15 kilometres or so from Newgrange. Here, the story of the Rockabill Islands, which consist of a large rock and a smaller one, says that they are the cow and calf which had been turned to stone when Balor opened his eye. It is no coincidence that there is a winter solstice alignment pointing at these islands from two standing stones at Baltray, which overlooks the mouth of the Boyne and the sea. The larger stone has a flat edge pointing directly at Rockabill. On the shortest day of the year, the sun rises from behind the Rockabill Islands, perhaps giving a cosmological explanation for the story of Balor and the cow and calf.

Bóinn may well be the personification of the moon, and the bright band of the Milky Way in the night sky could be connected with the bovine mythology. Certain versions of the magic cow myth describe her as the *Glas Ghoibhneann,* the cow that can fill every pail put under her with milk. In one story, a jealous woman claimed that the she possessed a vessel which the *Glas Ghoibhneann* could not fill. She brought a sieve and began milking the cow. The Glas continued to produce a continuous stream of milk, but it all ran through the sieve. Is this spray of milky droplets intended to represent the Milky Way in myth?

In old Ireland the name for the Milky Way galaxy was *Bealach/Bóthar na Bó Finne,* meaning the way/road of the white cow. Thus we can suggest that the Boyne River, which loops around the great monuments of Newgrange, Knowth and Dowth, may have been imagined as the earthly reflection of the great sky river. In *Island of the Setting Sun,* Richard Moore and I suggest that the milky quartz wall around the front of Newgrange may have been an effort to mimic the Milky Way, which at certain times of the year would have been visible as a giant ring sitting on the horizon from north to east to south to west and back to north again. It is in the pages of this book that we explore the idea, one of many, that the people who built the great monuments of the Boyne were seasoned astronomers, well acquainted with the movements of the sun, moon, planets and stars above their heads, and endowed with generations of experience of watching the sky. They used this knowledge to aid their grand construction projects, and indeed cosmology is intrinsic to the design of the mounds.

Newgrange is world famous for its Winter Solstice alignment. At dawn on the shortest days of the year, the light from the rising sun enters a specially constructed stone aperture above the entrance and penetrates down the narrow passageway into the central chamber. This achievement is hailed as an ancient triumph, and is accepted as being an intentional feature of New-grange's design.

But this alignment represents just one small part of a much more complex picture. At nearby Dowth, one of its two passageways admits light at sunset on Winter Solstice. At Cairn T in Loughcrew, some 25 miles northwest of the Boyne Valley monuments, sunrise on the spring and autumn equinoxes shines down the passage and illuminates a carved stone at the end of the central chamber.

But the ancient calendar was not based on solar movement alone. It would be cumbersome to construct a calendar based purely on the sun's move-ments. It was possible to have a much more precise and practical calendar based on a combination of solar, lunar, planetary and stellar movements. Time could be divided much more intricately by referring to the moon as well as the sun. It would have been obvious from the very earliest times that the moon passes through phases, and that it takes about 29 days for the moon to return to the same phase, i.e. full moon to full moon. It would also have been observable over a period of time that the moon returns to the same position against the background stars in less time, just 27 days. It would equally have been noticeable that the moon's path through the stars wanders somewhat from that of the sun.

While all this might seem to complicate the construction of a calendar, knowledge of how the sun and moon interact is essential to good time keep-ing. If, for instance, cloudy conditions have obscured the sunrises for a week around Winter Solstice, without knowledge of the lunar movements one cannot know precisely when the shortest day falls. But this becomes possible when one is armed with an understanding of the lunar and planetary move-ments. The full moon, for instance, is always located in the opposite part of the sky to the sun. In this way, it is possible to know which zodiac constella-tion the sun is in by looking at what constellation the full moon is in.

In addition to all this, it would have been perceptible over a long period of time that lunar eclipses, when the full moon darkens as a result of entering the earth's shadow, happened at predictable intervals.

A prolonged study of the Boyne monuments, their carvings, alignments and mythology, leads us to conclude that the people who built these great mounds had a high level of astronomical competency. They went far beyond the simple measurement of the year based on the sun's rising and setting positions on the horizon.

At Newgrange, for instance, the sun was not the only heavenly body which could cast its light into the chamber. Because the full moon is always opposite the sun, it follows that at the time of summer solstice, the longest days, the moon would be rising in the winter solstice position. An observer inside Newgrange would, given the right weather conditions, be able to see moonlight in the inner chamber at this time.

At nearby Knowth, there are many intricately carved kerbstones, including one aptly named the "Calendar Stone," so called because it appears to trace out calculations representing a five-year moon period closely tied in with a longer 19-year cycle of the moon known as the Metonic Cycle. The number of kerbstones at Knowth is 127, half of 254, the number of sidereal lunar months in a 19-year Metonic Cycle. At Dowth, the third of the great mounds, there are 115 kerbstones, half of 230, the number of synodic lunar months in a moon-swing cycle, the 18.6-year period during which the moon appears to "swing" between its extremes of declination.

The moon's movements, apparently complicated to us modern people who live indoors and who no longer watch the stars at night, would doubtless have been more obvious to the ancient mound builders who lived lives much more attuned with nature. But the sun and moon might not have been the only objects of interest to these ancient sky watchers.

Another object which coincidentally shared the sun's winter solstice declination in the Neolithic was Sirius, the Dog Star, the brightest star in the night sky. An astronomer hunched down in the chamber of Newgrange would be able to see this brilliant star shining in through the roofbox.

And that's not all. Another of the celestial bodies, the planet Venus, is known to rise in the morning before sunrise, and at other times to be visible in the evening sky in the west after sunset. Hence it has become known as both "The Morning Star" and "The Evening Star."

There is an old story in the Boyne Valley, recorded by the American writer Joseph Campbell, which says that the morning star cast a beam of light into the chamber of Newgrange. This folk tale says the light of Venus shone directly onto a stone containing two worn sockets, and that this phenomenon happened just once in eight years. Incredibly, Venus does have a cycle of eight years, after which its pattern of morning and evening apparitions appears to repeat itself again.

Christopher Knight and Robert Lomas, writing in *Uriel's Machine*, suggest that a series of eight rectangular markings on the stone marking the upper edge of the Newgrange roofbox represent the eight years of the Venus cycle. They further suggest that he roofbox was particularly designed to "trap" light from certain heavenly bodies, such as Venus, while also minimising scattered light from the sky.

Once in eight years, on the morning of the winter solstice, Venus would have risen over Red Mountain, an aptly named hill across the valley from Newgrange, and its light would have penetrated deep into the chamber. Some time later, the sun would rise and also cast its light into the deepest belly of the mound, much more brilliantly.

Venus is sometimes referred to in Irish as *caillichín na mochóirighe,* which means "early rising little hag." From this we might infer that the hag was the moon, the little hag Venus, and that the cow and calf of Irish myth are analogous with the hag and the little hag. The *cailleach* (hag) is a very ancient goddess who was venerated in various parts of Ireland in prehistoric times. The hills upon which the 6,000-year-old cairns of Loughcrew sit are called *Sliabh na Caillaighe,* the hills of the hag. To call this ancient goddess a hag seems derogatory, and indeed hag is just one translation of the word *cailleach* which can also mean "veiled one," "nun" and, interestingly, "virgin."

We have a compelling candidate in Bóann as one representation of this ancient *cailleach* goddess. The Boyne Valley is her landscape, and Newgrange itself represents her belly or womb. Indeed Brú na Bóinne, which the whole monument complex of the Boyne is called today, may be derived from the old Irish word *brú* which means, literally, "womb," but is more commonly believed to be derived from the root *brug,* which means "mansion" or "palace." Whatever the origin of the name Brú na Bóinne, Newgrange does possess very interesting design characteristics which can be compared with a female reproductive system.

Many researchers and authors have pointed out these unique aspects of the design of Newgrange. The entrance represents the vulva, the passage is the birth canal, and the chamber the uterus. Knight and Lomas believe this was a deliberate intention of the design, while historian William Battersby said the whole ridge upon which Newgrange sits presents "the impression of a large image of a pregnant woman." Newgrange might have been constructed to represent both a tomb and a womb. Archaeologist Marija Gimbutas said that throughout old Europe such monuments represented both tombs and shrines, and that the structures often took the shape of the female body.

These giant edifices were constructed not merely so that the bones of a select contingent of the ancient population could be interred within. They were rather seen as places of rebirth, where the soul of the deceased person could be transported into the otherworld, often seen as being located among the stars.

In one story, called "The Pursuit of Diarmaid and Gráinne," from the Ossianic Cycle of tales, we are told that the deceased hero Diarmaid was taken to Newgrange by Aonghus, a god of the Tuatha Dé Danann, to "put an aerial life into him so that he will talk to me every day." Aonghus "and since I cannot restore him to life I will send a soul into him, so that he may talk to me each day." In this story we might see a profound explanation for the true purpose for which Newgrange was built.

Aonghus was the supernatural offspring of divine parents. He is often referred to in myth as *Mac Ind Oc* or *Mac an Óg,* commonly held to mean "son of the young" but which has another, more intriguing, meaning. He is the son of the Dagda, the chief god of the Tuatha Dé Danann, and Bóinn, although Bóinn was married to Elcmar. According to myth, Dagda sent Elcmar on a journey and cast a magic spell so that the night disappeared and nine months seemed like just a single day and night. While Elcmar was away, Dagda slept with Bóinn and she bore him a child, Aonghus, the *Mac Óg*, so called because he was "begotten at the beginning of a day and born between that and the evening." In this context, it would seem apt to suggest that Bóinn is *cailleach* as "virgin," and that the miraculous story of Aonghus being born in a single day might be a prehistoric immaculate conception of sorts. Another Irish word for virgin is *Óige*, or *óigbhean*, meaning "young" or "young woman," a polite and dignified way of referring to a virgin.

Another miraculous virgin birth which takes place at Newgrange involves Setanta, later to become known as Cúchulainn, the hero of the Táin Bó Cuailnge epic, and a supreme champion featured in many Irish myths and stories. In this instance, Dechtine comes to Newgrange while chasing swans away from Emain Macha, the plain of Armagh. There she is visited by Lugh, another of the Tuatha Dé Danann, who tells her in a dream that she will bear a son, to be named Setanta.

Swans feature strongly in the story of Aonghus, who falls in love with a maiden whom he had seen in a dream. After a prolonged search of Ireland, involving some of the Tuatha Dé Danann, the object of his desires is found in Tipperary. She is revealed as Caer Iobharmhéith, who is from Sídh Uamhain, an otherworld residence in Connacht, and when she is found she has taken the form of a swan by a lakeside.

When Aonghus went to Caer, he was miraculously transformed into a swan. The two embraced and flew three times around the lake before flying together to Brú na Bóinne where they "put the dwellers of that place to sleep with their beautiful singing." The story says that they remained in the Brú after that.

The prominent myths featuring swans at Newgrange are interesting because of the long-term presence in the Bend of the Boyne of the Whooper Swan, which gathers here every winter in large numbers. The passage and chamber of Newgrange are cruciform, bringing to mind the cross-shaped swan constellation of the night sky, which we know today as Cygnus.

Of further interest is the fact that the axis of the passage of Newgrange points to a smaller passage-mound called Fourknocks, located on a hilltop some 15 kilometres to the southeast. Fourknocks in turn has a short passageway and egg-shaped chamber which is aligned towards the far northeast, well beyond the range where the sun or moon could possibly shine into its bosom. One target, if there was one, of the Fourknocks chamber, might have been Deneb, the bright star of the swan constellation, which would have been rising in the far northeast at the time Fourknocks was constructed.

If Newgrange and Fourknocks represent some concern with the swans which visited the Boyne, and the stellar swan constellation, perhaps this is related to precession of the equinoxes, the slow spinning of the axis of the earth over a period of 25,800 years. This effect, the product of a wobble in the earth's axis, causes the positions of the stars to drift over time as seen from the surface of the earth. Two of the most noticeable effects of precession are the changing of the star that marks the northern pole of the sky, and the turning of the Zodiac, causing the sun's solstice and equinox positions to drift westwards over long periods of time. At the time Newgrange was built, the summer solstice sun was in the constellation we know today as Leo. Today, the summer solstice sun is located between Taurus and Gemini, above the upraised hand of Orion.

The significance of Deneb at the time Newgrange was built relates to the fact that this was the only time in the entire precessional cycle when Deneb would have been seen to set below the horizon. At all other times throughout the long 25,800-year cycle, Deneb would be what we call "circumpolar" meaning it is visible throughout the night and neither rises nor sets. There is one century-long period, perhaps a bit more, in this long cycle when Deneb swings down low enough to hit the horizon, and this coincided with the time Newgrange was built.

The cross, representing the swan, and perhaps the swan-god Aonghus, the lover of Caer and son of the great goddess Bóinn, is an interesting symbol. We are familiar today with the cross as the symbol of Christ, the divine child born of a virgin as a result of an immaculate conception. But in the Boyne Valley we see that the cross has a much more ancient origin, predating Christianity by more than three millennia. In ancient times, it also represented the divine child, Aonghus, supernatural son of a virgin mother, Bóinn, who was venerated at Newgrange and in the Boyne Valley in far-off times.

She was a stellar, lunar and earth goddess, who represented birth, the cycle of life, and the miraculous transformation of the soul and its journey to the afterlife. Newgrange was her womb, designed in part as a rebirth chamber where the soul of the dead could be assisted in its journey to a stellar otherworld. She was the archetypal *mater dei,* part of an ancient divine trinity including the Dagda, the supreme god, and Aonghus, the miracle swan child who had taken ownership of Newgrange from his father.

And on the shortest days of the year, at dawn, a shaft of light from the rising sun comes into the dark womb and there the year is born again. Once every eight years, the *caillichín na mochóirighe,* Venus, shines into the chamber some time before the sun. An even rarer conjunction, perhaps the one most celebrated at Newgrange, involved the meeting of the last crescent moon beside the morning star, hanging together in the dawn sky on the shortest day before sunrise. Is this the meaning of the triple spiral symbols found on the main entrance kerbstone at Newgrange and a stone in the rear of the chamber? Does this tri-spiral symbol represent the meeting of sun, moon and morning star? There is no doubt that the moon and Venus, the hag and the little hag, were watched carefully in the Boyne Valley in remote times, and that at least some information about their movements were engraved into stone and encoded into myth.

Appendix I

The Astronomy of the Nights of Venus

Barbara Carter

N INIS R NIGHTS

This chart shows the nights of the calendar that are marked with N INIS R over all 5 years, which we have followed to be Venus. In the ancient Gaulish language inis means island, so we have come to think of Venus as the Island in the sky. The calendar, as a general rule, notes Venus three nights before the full Moon and three nights after the full Moon, and three nights before the new Moon, on the new Moon, and three nights after the new Moon. Two months in the fall, Eqvos and Edrinios, have no notes to Venus. The Full and New Moons generally count to the VIII in both the first fifteen nights and Atenovx VIII in the last fourteen or fifteen nights.

There are many abbreviations in this chart:

To next count, elong., Month, Day, date, Astro, A/PM, time, sunlight, AZM, and East/West.

Elong. = elongation is the angle between the Sun and a planet, or between a plant and a satellite, as seen from the Earth. Elongation is measured along the ecliptic in degrees west or east of the Sun.[1] In regard to this also, Ptolemy had used what he called *arcus visionis,* of thirteen degrees for the Moon. "Only when the Sun was 13 degrees below the horizon was the sky dark enough for a crescent Moon to be seen on the horizon."[2]

The months of the Sequani/Cologny Calendar are the Intercalary month called Ogmios or Antaran, Samonios, Dvmannios, Rivros, Anagantios, Ogronios, Cvtios, Giamonios, Simivisonn, Eqvos, Elembivios, Edrinios, Cantlos. The days of the months are numbered in Roman numerals with the first half of the month (from first quarter Moon to third quarter Moon) numbered from I to XV; the second half of the month (from third quarter Moon to first quarter Moon) is labeled ATENOVX which we think means last nights. This has been abbreviated to ATE. Since the months can contain thirty or twenty-nine nights, these last nights are numbered I to XV or I to XIIII.

The months of the Gregorian calendar are all abbreviated to three letters, so this is all of them exempt May. Astrology is abbreviated to Astro. All the signs of Astrology have been abbreviated to two or three letters. These signs are Aries, Taurus, Gemini, Cancer, Leo, Virgo, Libra, Scorpio, Sagittarius, Capricorn, Aquarius, Pisces. Retrograde is abbreviated RX.

The rings mentioned in this charting are the three rings of sunlight, two of which shine before the Sun rises and after the Sun sets. These are shown on the older versions of sky maps produced for computers. The first ring of sunlight is the one sun occupies. Anthony Murphy tells the story of Balor in ancient Ireland, with his single eye in the middle of his forehead. There were always seven coverings over his eye.[3] There were many times in this charting that we wished to be able to explain how many covering were over Balor eye to explain the stages of dusk, twilight, and dawn in relation to the Sun and Venus. However, the rings of the Sun have been helpful, which are not usually shown on the latest versions of star maps for computers. The words up and high have been used to describe Venus being away from the Sun. High is more; while up is less. Both are in the twilight. Horizon has been abbreviated to hor.

AZM is the abbreviation used for *azimuthal coordinates,* which are also called horizontal coordinates. This is the simplest positional system. Their reference plane is the observer's horizon, and the object's position in terms of altitude and azimuth (or zenith distance). The azimuthal coordinates in this charting has used the reference place of Elkins, West Virginia.

The directions have also been abbreviated.

The following chart shows the N INIS R Nights over 5 years. All numbers relate to Venus. Other planets have been included.

#	to next count	elong.	Month	Day	date	Astro	A/PM	time	sunlight	AZM	east/west
1.	18 nights	4.7	SAM	VII N DVMANN INIS R	Dec. 28, 2001	2Cap	AM	7:47	1st ring	125°03'	SE
2.	17 nights	.90	SAM ATE	VIIII N DVMANNI IN R	Jan. 14, 2002	23Cap	PM	7:47	same	as Sun	SE
3.	6 nights	4.39	DVM	XI N INIS R	Jan. 31, 2002	14Cap	PM	5:22	1st ring	43°05'	SW
4.	7 nights	9.7	DVM ATE	XII N INIS R IVOS	Feb. 16, 2002	4Pis	PM	5:54	2nd ring	50°47'	SW
5.	6 nights	9.85	RIV	V D INIS R	Feb. 23, 2002	13Pis	PM	6:14	end 2nd	55°15	SSW
6.	11 nights	11.29	RIV	XI N INIS R	Mar. 1, 2002	21Pis	PM	6:17	end 2nd	257°52'	SWW
7.	2 nights	13.25	RIV ATE	VII N ANAG INIS R	Mar. 12, 2002	4Ari	PM	6:22	twilight	261°42'	W
8.	11 nights	14.44	RIV ATE	VIIII N ANAG INIS R	Mar. 14, 2002	same	same	same	2nd ring	same	
9.	6 nights	17.11	ANAG	V N INIS R	Mar. 25. 2002	7Ari	PM	6:42	twilight	269°14'	NW
10.	13 nights	21.04	ANAG ATE	VI N INIS R	Apr. 10, 2002	10Ta	PM	6:57	up	275°47'	NofW
11.	4 nights	24:23	OGRO	V N INIS R	Apr. 23, 2002	26Ta	PM	7:22	twilight	281°42'	NW
12.	18 nights	25.22	OGRO	VIIII N CVTIO INIS R	Apr. 27, 2002	1Ge	PM	7:23	up	282°38'	NW
13.	9 nights	29.60	OGRO ATE	XII N INIS R	May 15, 2002	23Ge	PM	7:34	twilight	283°31'	N&W
14.	4 nights	31.50	CVT	V N INIS R	May 23, 2002	2Ca	PM	7:34	twilight	282°16	NWW
15.	10 nights	32.34	CVT	VIIII N GIAMO INIS R	May 27, 2002	7Ca	PM	7:34	twilight	281°40'	W
16.	2 nights	34.73	CVT ATE	IIII N INIS R	Jun. 6, 2002	19Ca	PM	7:49	after sunset	280°52'	W
17.	1 night	35.18	CVT ATE	VI N INIS R	Jun. 8, 2002	21Ca	PM	same	as above	same	same
18.	1 night	35.40	CVT ATE	VII N GIAMO INIS R	Jun. 9, 2002	23Ca	PM	same	as above	same	same
19.	33 nights	35.62	CVT ATE	VIII N GIAMO INIS R	Jun. 10, 2002	24Ca	PM	same	high	same	NW
20.	8 nights	42.13	GIAM ATE	XI N INIS R	Jul. 13, 2002	2Vi	PM	7:37	daylight	263°52'	W
21.	65 nights	43.37	SIMIVI	V N INIS R	Jul. 21, 2002	11Vi	PM	7:43	sunset	261°27'	due W
22.	68 nights	39.91	ELEM	X N INIS R	Sep. 24, 2002	10Sc	PM	on hor.	at twilight	236°00'	SSW
23.	33 nights	37.09	CANTL ATE	IIII N INIS R	Dec. 1, 2002	1Sc	AM	7:04	dawn	122°52'	SE
24.	4 nights	46.75	SAM ATE	VIII N INIS R	Jan. 3, 2003	25Sc	AM	7:04	dawn	145°46'	SE
25.	13 nights	46.80	SAM ATE	VIIII N INIS R	Jan. 4, 2003	26Sc	AM	7:04	dawn	146°00'	SE
26.	4 nights	46.80	DVM	VII N INIS R	Jan. 17, 2003	9Sa	AM	7:02	dawn	147°02'	SE
27.	13 nights	46.63	DVM	XI N INIS R	Jan. 21, 2003	13Sa	AM	same	same	147°33'	SE
28.	3 nights	45.49	DVM ATE	VIIII N INIS R	Feb. 3, 2003	28Sa	AM	6:02	dawn	134°02'	SE
29.	7 nights	45.14	DVM ATE	XII N INIS R	Feb. 6, 2003	1Cap	AM	same	same	133°09'	due SE

30.	6 nights	44.42	RIV	V N INIS R	Feb. 13, 2003	9Cap	AM	6:33	dawn	137°44'	SE
31.	24 nights	43.34	RIV	XI N INIS R	Feb. 19, 2003	16Cap	AM	6:29	same	36°29'	SE
32.	17 nights	39.14	ANAG	V N INIS R	Mar. 15, 2003	14Aq	AM	5:37	dawn	118°27'	SE
33.	1 night	35.89	ANAG ATE	VI N INIS R	Mar. 31, 2003	2Pis	AM	5:37	dawn	113°36'	SE
34.	2 nights	35.68	ANAG ATE	VII N INIS R	Apr. 1, 2003	5Pis	AM	5:50	on hor.	107°10'	SE
35.	10 nights	35.25	ANAG ATE	VIIII N INIS R	Apr. 3, 2003	11Pis	AM	5:51	on hor.	108°59'	SE
36.	23 nights	33.06	OGRO	V N INIS R	Apr.13, 2003	19Pis	AM	5:12	dawn on hor.	102°40'	SE
37.	9 nights	27.29	OGRO ATE	XII N INIS R	May 5, 2003	16Ar	AM	4:37	dawn	87°26'	due E
38.	5 nights	25.98	CVT	V N INI R	May 13, 2003	25Ar	AM	4:40	dawn	84°20'	NofE
39.	11 nights	24.98	CVT	VIIII N INI R	May 17, 2003	0Ta	AM	4:40	dawn	84°11'	dueE
40.	3 nights	22.45	CVT ATE	IIII N INI R	May 27, 2003	20Ta	AM	4:43	dawn	78°50'	NofE
41.	3 nights	21.94	CVT ATE	VI N INI R	May 29, 2003	22Ta	AM	same	w/ Moon	78°02'	NE
42.	19 nights	17.34	GIAM	VIIII N INIS R	Jun. 16, 2003	7Ge	AM	4:50	dawn	71°31'	midNE
43.	1 nights	13.17	GIAM ATE	VII N INI R	Jun. 29, 2003	22Ge	AM	4:53	dawn	67°52'	NNE
44.	4 nights	13.50	GIAM ATE	VIII N INI R	Jun. 30, 2003	24Ge	AM	4:54	dawn	68°37'	NE
45.	9 nights	12.69	GIAM ATE	XI N INI R	Jul. 3, 2003	27Ge	AM	4:51	dawn	66°43'	NE
46.	63 nights	10.53	SIMI	V N INIS R	Jul. 11, 2003	7Ca	AM	4:57	dawn	65°45'	NE
47.	69 nights	7.01	ELEM	X N INI R	Sep. 12, 2003	25Vi	PM	6:27	in sunlight	268°15'	W

48.	34 nights	24.03	CANT ATE	IIII N INIS R	Nov. 19, 2003	20Sa	PM	5:01	twilight	225°06'	SW
49.	1 night	31.45	SAM ATE	VIII N INIS R	Dec. 22, 2003	0Aq	PM	5:16	twilight	220°02'	SW
50.	14 nights	31.67	SAMO ATE	VIIII N INIS R	Dec. 23, 2003	2Aq	PM	5:00	twilight	216°22'	SSW
51.	14 nights	34.39	DVM	VII N INIS R	Jan 5, 2004	16Aq	PM	5:14	twilight	219°59'	dueSW
52.	4 nights	37.76	DVM ATE	VIIII N INIS R	Jan. 22, 2004	8Pis	PM	5:42	twilight	27°33'	dueSW
53.	4 nights	38.33	DVM ATE	XII N INIS R	Jan 25, 2004	12Pis	PM	5:44	twilight	228°47'	SW
54.	8 nights	39.60	RIV	V N INIS R	Feb 1, 2004	20Pis	PM	5:46	twilight	231°11'	SW
55.	25 nights	40.64	RIV	XI N INIS R	Feb. 7, 2004	28Pis	PM	5:46	twilight	233°30'	SW
56.	17 nights	44.18	ANAG	V D INIS R	Mar. 2, 2004	25Ar	PM	6:51	twilight	256°24'	W
57.	1 night	45.66	ANAG ATE	VI N INIS R	Mar. 18, 2004	13Ta	PM	6:55	twilight	263°09'	W
58.	1 night	45.72	ANAG ATE	VII N INIS R	Mar. 19, 2004	14Ta	PM	6:58	twilight	263°35'	W
59,	3 nights	45.56	ANAG ATE	VIIII N INIS R	Mar. 21, 2004	16Ta	PM	7:11	twilight	266°57'	veryW
60.	22 nights	45.98	OGRO	V N INIS R	Mar. 31, 2004	26Ta	PM	7:15	twilight	270°45'	NWW
61.	9 nights	43.21	OGRO ATE	XII N INIS R	Apr, 22, 2004	15Ge	PM	8:26	twilight	286°09'	NWW
62.	5 nights	40.44	CVT	V N INI	Apr. 30, 2004	20Ge	PM	8:22	twilight	287°39'	NWW
63.	11 nights	30.52	CVT	VIIII N INI	May 4, 2004	22Ge	exactly	same	as above	same	same
64.	3 nights	31.62	CVT ATE	IIII N INI R	May 14, 2004	25Ge	PM	8:01	twilight	288°50'	NWW
65.	19 nights	29.86	CVT ATE	VI N INI R	May 16, 2004	26Ge	PM	8:06	twilight	290°06'	NWW

66.	1 night	6.87	OGM	VIIII NO GIAMON INIS R EQVI	Jun. 3, 2004	21Ge	PM	7:35	2nd ring	296°33'	NWW
67.	46 nights	5.6	OGM	X N ELEM INIS R	Jun. 4, 2004	20Ge	PM	7:41	1st ring	297°41'	NWW
68.	9 nights	41.40	GIAM ATE	XI N INI R	Jul. 19, 2004	15Ge	AM	4:52	dawn	88°56'	dueE
69.	66 nights	43.95	SIM	IV V D INIS R	Jul. 28, 2004	20Ge	AM	5:02	dawn	90°46'	due E
70.	69 nights	41.22	ELEM	X N INIS R	Oct. 1, 2004	26Leo	AM	5:59	dawn	104°40'	veryE
71.	5 nights	27.03	CANT ATE	IIII N INI R	Dec. 8, 2004	14Sc	AM	6:00	dawn	127°27'	SE
72.	1 night	26.11	CANT ATE	VIII N SAMON INIS R	Dec. 12, 2004	24Sc	AM	6:15	dawn	121°15'	SEE
73.	13 nights	25.88	CANT ATE	VIIII N SAMON INIS R	Dec. 13, 2004	25Sc	AM	6:15	dawn	132°09'	SE
74.	18 nights	23.10	SAM	VII N DVMANN INIS R	Dec. 25, 2004	0Sa	AM	7:15	dawn	132°06'	dueSE
75.	18 nights	33.2	SAM ATE	VIIII N DVM.INIS.R	Jan. 11, 2005	1Cap	AM	7:27	dawn	129°23'	dueSE
76.	8 nights	15.12	DVM	XI N INIS R	Jan. 28, 2005	20Sag	AM	7:20	dawn	125°18'	SE
77.	17 nights	11.34	DVM ATE	XII N INIS R IVOS	Feb,13, 2005	12Aq	AM	7:13	dawn	116°15'	SE
78.	8 nights	9.60	RIV	V D INIS R	Feb. 20, 2005	21Aq	AM	6:57	dawn	111°36'	SEE
79.	7 nights	8.25	RIV	XI N INIS R	Feb. 26, 2005	29Aq	AM	6:54	dawn	108°18'	SE
80.	12 nights	5.62	RIV ATE	VII N ANAG INIS R	Mar. 9, 2005	12Pis	AM	6:36	sunrise	100°08'	SEE

81.	3 nights	5.15	RIV ATE	VIIII N ANAG INIS R	Mar. 11, 2005	15Pis	AM	6:42	sunrise	100°13'	SEE
82.	12 nights	2.62	ANAG	V N INIS R	Mar. 22, 2005	29Pis	AM	6:17	sunlight	87°29'	E
83.	14 nights	2.28	ANAG ATE	VI N INIS	Apr. 7, 2005	20Ar	PM	6:28	sunset	275°26'	NofW
84.	14 nights	5.43	OGR	V N INIS R	Apr. 20, 2005	5Ta	PM	6:52	daylight	282°23'	NW
85.	5 nights	6.44	OGR	VIIII N QVTIO INIS R	Apr. 24, 2005	24Ca	PM	6:54	daylight	282°34''	NWW
86.	19 nights	11.14	OGR ATE	XII N INIS R	May 12, 2005	2Ge	PM	7:22	twilight	290°96'	NWW
87.	9 nights	13.26	CVT	V N INIS R	May 20, 2005	12Ge	PM	7:24	twilight	290°14'	NWW
88.	5 nights	14.33	CVT	VIII N GIAMO INIS R	May 24, 2005	16Ge	PM	7:30	twilight	290:51	NWW
89.	10 nights	17.00	CVT ATE	IIII N INIS R	Jun. 3, 2005	27Ge	PM	7:40	twilight	290:50	NW
90.	11 nights	17.53	CVT ATE	VI N INIS R	Jun. 5, 2005	1Ca	PM	7:31	daylight	289:03	NWW
91.	1 night	17.80	CVT ATE	VII N GIAMO INIS R	Jun. 6, 2005	2Ca	PM	7:34	daylight	289°36'	NWW
92.	1 night	18.07	CVT ATE	VIII N GIAMO INIS R	Jun. 7, 2005	4Ca	PM	7:36	daylight	289°40'	NWW
93.	17 nights	22.31	GIAM	VIIII N INIS R	Jun. 23, 2005	23Ca	PM	7:50	twilight	287°18'	NWW
94.	14 nights	25.70	GIAM ATE	VII N INI R	Jul. 6, 2005	9Le	PM	7:54	twilight	282°44'	W
95.	1 night	25.95	GIAM ATE	VIII N INIS R	Jul. 7, 2005	10Le	PM	7:55	twilight	282°28'	W

96.	4 nights	26.72	GIAM ATE	XI N INI R	Jul. 10, 2005	14Le	PM	7:57	twilight	281°29'	W
97.	9 nights	28.75	SIM	V N INIS R	Jul. 18, 2005	23Le	PM	8:00	twilight	278°14'	W
98.	64 nights	42.65	ELE	X N INI R	Sep. 20, 2005	10Sc	PM	6:44	twilight	237°45'	W
99.	69 nights	44.61	CANT ATE	IIII N INIS R	Nov. 26, 2005	18Cap	PM	5:02	twilight	205°03'	S
100.	1 night	22.47	SAM	VIII N INIS R	Dec. 29, 2005	1AqRx	PM	6:01	twilight	226°59	SW
101.	14 nights	21.22	SAM	VIIII N INIS R	Dec. 30, 2005	1AqRx	PM	6:01	twilight	238°58'	SW
102.	5 nights	5.57	DVM	VII N INIS R	Jan. 12, 2006	24CpRx	PM	5:02	daylight	244°14'	SW
103.	14 nights	7.69	DVM	XI N INIS R	Jan. 16, 2006	22CpRx	AM	7:20	dawn	13°57'	SE
104.	14 nights	23.90	DVM ATE	VIIII N INIS R	Jan. 29, 2006	16CpRx	AM	6:24	dawn	11°:49'	SE
105.	4 nights	27.26	DVM ATE	XII N INI R	Feb. 1, 2006	16Cp	AM	6:28	dawn	119°44'	SSE
106.	8 nights	33.65	RIV	V N INIS R	Feb. 8, 2006	16Cpdirect	AM	6:10	dawn	121°22'	SE
107.	7 nights	37.76	RIV	XI N INIS R	Feb. 14, 2006	17Cp	AM	6:00	dawn	123.°06'	SE
108.	25 nights	45.62	ANAG	V N INI R	Mar. 10, 2006	13Aqdir.	AM	6:05	dawn	128°53'	dueSE
109.	17 nights	46.53	ANAG ATE	VI N INI R	Mar. 26, 2006	19Aq	AM	5:46	dawn	124°00'	SE
110.	1 night	46.51	ANAG ATE	VII N INI R	Mar. 27, 2006	19Aq	same	as above	dawn	124°08'	SE
111.	3 nights	46.48	ANAG ATE	VIIII N INI R	Mar. 29, 2006	21Aq	same	as above	dawn	121°02'	SE
112.	11 nights	45.94	OGRON	V N INIS R	Apr. 8, 2006	29Aq	AM	5:30	dawn	118°49'	SE
113.	23 nights	43.49	OGRON ATE	XII N INIS R	Apr. 29, 2006	23Pis	AM	4:24	dawn	100°21'	E

114.	9 nights	42.22	CVT	V N INIS R	May 7, 2006	4Ar	AM	4:30	dawn	98°00'	E
115.	11 nights	41.54	CVT	VIIII N INI R	May 11, 2006	8Ar	AM	4:30	dawn	97°08'	E
116.	3 nights	39.69	CVT ATE	IIII N INI R	May 21, 2006	20Ar	AM	4:30	dawn	94°26	dueE
117.	19 nights	39.29	CVT ATE	VI N INI R	May 24, 2006	22Ar	AM	4:30	dawn	91°49'	dueE
118.	14 nights	35.31	GIAM	VIIII N IMI R	Jun. 11, 2006	13Ta	AM	4:20	dawn	82°46'	E
119.	1 night	32.34	GIAM ATE	VII N INIS R	Jun. 24,2006	28Ta	AM	4:05	dawn	75°49'	NE
120.	4 night	32.10	GIAM ATE	VIII N INIS R	Jun. 25, 2006	29Ta	AM	4:30	dawn	78°57'	NE
121.	9 nights	31.39	GIAM ATE	XI N INIS R	Jun. 28, 2006	3Ge	AM	4:36	dawn	78°57'	NE
122.	66 nights	29.47	SIM	V N INIS R	Jul. 6, 2006	13Ge	AM	4:26	dawn	74°34'	NW
123.	69 nights	12.74	ELEM	X N INIS R	Sep. 9, 2006	2Vi	AM	5:07	dawn	77°35'	NEE
124.	29 nights	4.85	CANT ATE	IIII N INIS R	Nov. 15, 2006	27Sc	PM	5:00	twilight	242°27'	SW
125.	1 night	12.85	OGM ATE	VII NSDS SAMONI ANAGAN INNIS TIT	Dec. 18, 2006	7Cp	PM	5:00	twilight	231°06'	SW
126.	15 nights	13.08	OGM ATE	VIII NSDS TO INN	Dec 19, 2006	8Cp	PM	same	as last	231°08'	SW

NOTES FOR CHART

8. W/ New Moon.

11. Venus w/ Pleiades. Mars close to Mercury.

12. Saturn, Mars, Pleiades, Venus, Mercury.

13. Venus @ the end of Horns of the Bull. Mars, Saturn in Horns.

14. Line up: Jupiter, Venus, Mars. Saturn, Mercury. Sun w/Pleiades.

16. Venus w/ Castor/Pollux & Jupiter. Sun in Horns of Bull.

17. Venus high w/ Castor/Pollux.
18. Almost new Moon.
19. W/Pollux. New Moon.
20. W/Regulus & crescent Moon.
21. From AM 239 nights. W/Regulus.
22. From AM 307 nights.
23. 340 nights from AM — 23 counts. Venus, Mars, crescent Moon, Spica.
24. Begin year 2. Venus, Mars high between Spica & Antares.
25. Venus moving from Mars.
26. Venus w/ Antares.
28. Venus middle of Mars and Mercury.
30. Venus on top Sagittarius. Mars w/ Antares.
31. Ecliptic crossing East/West.
32. Venus on hor. sunrise. Venus rise- Moon set.
33. On Mar 29, w/ Moon, Sun, Venus. Moon below Equuleus.
37. Rises right before Sun.
38. Notation from year 5. Rises right before Sun.
39. Rises before Sun. Mercury 3rd ring.
40. Rises w/ Mercury, Pleiades before Sun.
42. Sun enters Horn of Bull. Venus top Aldebaran.
43. Venus 3rd ring. Saturn and Moon 1st ring. Mercury 2nd ring.
44. Venus in Horns of Bull. Mercury and Saturn w/ Moon on hor.
46. 23 counts of AM 223 nights. Venus w/ Saturn in 3rd ring of Sun.
48. 25 counts this year.
49. Year 3. Mercury on hor.
50. Dec 25, crescent Moon w/ Mercury on hor.
52. Moon on hor.
53. Moon 4 days old. Moon just passed Venus
55. Venus on intersection of the celestial equator and ecliptic.
56. Up from hor. Mars between Pleiades & Venus.
57. Close to maximum elongation.
58. Mars w/ Pleiades setting twilight w/ Jupiter rising.
59. Mars w/ Pleiades. Jupiter rising. New Moon on hor. setting.
60. Mars w/ Aldebaran. Venus w/Pleiades. Mercury on hor.
61. Nice lineup: Venus mid Mars &Moon. Saturn above Pleiades.
62. Recopied from year 5. Venus w/Mars in the Horns of the Bull.
63. Venus at the end of Horn of Bull. Full Moon.
64. Saturn close to Mars.
65. Venus almost retrograde. Mercury w/ Moon AM.
66. NO could be begin Venus calendar. Venus on hor.
67. 21 counts PM. 267 nights. Venus on hor. very close to Sun. On Jun. 8, Venus passed the face of Sun.

68. Venus conjunct to Aldebaran on Jul. 4.

69. Up from Sun outside the Horns of Bull.

70. Venus very high w/ Regulus.

71. Jupiter above Moon. Saturn w/ Castor & Pollux.

72. On Dec 9, Venus & Mars w/ old Moon E. Pleiades setting W.

73. 25 counts this year. Venus close to Mars. Horns of Bull setting.

74. Year 4. Venus rising w/ Mercury & Antares. Full Moon setting in Horns of Bull.

75. Venus top of Sagittarius w/ Mercury on hor.

76. Mercury & Venus 3rd ring on hor. Jupiter w/ Spica.

77. Mercury w/ Sun. Venus 2nd ring hidden by sunlight.

78. Venus 2nd ring right on hor. Mars top Sagittarius.

79. Venus hidden by Sun. Venus 2nd ring on hor.

80. Venus, crescent Moon, Sun rise together .

81. Venus hidden by Sun. Venus on hor. w/ Sun.

82. 15 counts AM— 246 nights. Venus w/ Sun. Mercury beside Jupiter sets as Sun rises.

83. Venus on hor. at sunset hidden by Sun. Moon/Mercury AM.

84. Venus hidden by Sun. Venus in 2nd ring.

85. Venus hidden by Sun. Venus on hor. 2nd ring as Sun sets.

86. Venus between Pleiades & Aldebaran- all hidden by Sun.

87. Copied from year 1. Venus in Horns of Bull 3rd ring.

88. Venus sets shortly after Moon rises. Venus 3rd ring of Sun.

89. Venus 3rd ring Saturn w/ Castor &Pollux.

90. Mercury on hor. Venus above Moon w/ Pleiades dawn.

91. New Moon w/ Mercury hor. Venus up w/ Sun in Horn of Bull.

92. Venus w/ crescent Moon on hor. Sun set in Horns of Bull.

93. Saturn, Venus, Mercury above 3rd ring w/ Castor & Pollux.

94. Venus & Mercury above Saturn & crescent Moon on hor..

95. Moon between Saturn & Venus w/Mercury.

96. Moon, Venus, Mercury above Saturn on hor.

97. Venus w/ Regulus/Mercury hor. Moon w/ Antares.

98. Venus halfway between Spica & Antares Jupiter w/ Spica hor.

99. 26 counts this year. Venus high w/ Sagittarius. Mars close to Pleiades rising.

100. Year 5. Venus on hor. out of Sun 2 hours from Venus set, Saturn rises.

102. 20 counts PM 281 nights. Venus on hor. in light of Sun 1st ring.

103. Venus on hor. in light of Sun 2nd ring.

104. Venus up out of light of Sun. Saturn just set.

105. Venus w/ Sagittarius up out of light of Sun.

106. Venus up out of light of Sun. Moon just past Aldebaran.

107. Venus up in dark 5:00 AM. Moon setting.

108. Venus far from Sun, rises 4:30 AM. Moon w/ Saturn.

109. Venus about as far from Sun as possible. Moon passing Venus.

111. Moon w/ Venus above Formalhaut under hor.

112. Venus & Mercury very high before Sun at dawn.

113. Venus very far from Sun on hor. Jupiter setting at dawn.

114. Venus very far from Sun on hor. Jupiter sets @ dawn.

115. Venus very far from Sun. As Venus rises Moon sets 3:34AM

116. 116.x 5=580.

117. 117x5=585. Moon passing Venus. Capella rising.

118. Venus, Pleiades,Capella at dawn rising out of Sun. Sun in Horns of Bull.

119. Venus w/ Pleiades.

120. Venus between Pleiades & Aldebaran. New Moon

121. Venus between Pleiades and Aldebaran. Capella at both dusk and dawn.

122. Venus in the Horns of the Bull.

123. 21 counts AM — 237 nights. Venus in 3rd ring of sunlight w/ Regulus, Saturn above.

124. 25 counts this year. 124 counts for 5 years. Venus w/Jupiter 1st ring of Sun setting w/ Antares.

In the first Intercalary month there are many nights marked with IN but only two that are very clearly related to Innis. This month has been named by the Sequani Calendar Studies Group as Ogmios yet may be labeled ANTA-RAN on the second line of this month. These nights are often labeled over two lines in the original.

125. First year. First Intercalary. Second cycle. 125. Venus just into 3rd ring of Sun 4 nights from Moon passing.

126. Intercalary night. New Moon. Geminid & Ursid Meteor showers. At dawn, Deneb rises, Coma Berenices close to zenith, Capella sets. Venus on top of Sagittarius on hor. at nightfall. Twilight Saturn rises w/ Regulus as Pleiades almost to zenith.

NOTES

1. Ridpath, Ian editor. *Norton ' s Star Atlas and Reference Handbook.* Essex, England: Addison Wesley Longman Limited, 19th edition 1998.

2. Hostetter, Clyde. *Star Trek To Hawa-i ' i: Mesopotamia To Polynesia,* San Luis Obispo: the Diamond Press, 1991.

3. Murphy, Anthony and Richand Moore. *Island of the Setting Sun: In Search of Ireland ' s Ancient Astronomers.* Dublin: the Liffey Press Ltd, 2006.

Appendix II

The Eight Year Cycle of Venus

Barbara Carter

GUIDE TO APPENDIX

We have come to call these strange marks on the calendar "the T marks." They appear in sets, but not always. The sets appear in threes: +||, |+|, ||+, but not always.

The two intersections of the galactic equator and the ecliptic are the places where the Sun crosses at the Solstices. The intersections of the celestial equator and the ecliptic are the places where the Sun crosses at the equinoxes. What we are able to say from this chart, is that this calendar keeps careful track of the Moon and the planets moving through these places in the sky. These can show the Major and minor standstills of the Moon. These can also show where Venus crosses the Sun both across the face of the Sun and behind the Sun changing from evening to morning star.

The months of the calendar have been abbreviated. The months are Samonios, Dvmannios, Rivros, Anagantios, Ogronios, Cvtios, Giamonios, Simivisonn.

*6 nights brought Moon to CE+E so these + marks also show how long it takes the Moon to move from GE+E 180 to GE+E 0.

1. SAM III +|| D EXINGIDVM IVOS Venus ready to cross the GE+E 0 Sun just crossed the GE+E 0	Year 1 December 24, 2001 eclipse new 12.30
2. SAM ATE VI ||+ M D Venus crosses the Sun Moon at the GE+E 0 Moon 28 nights	January 11, 2002
2a. SAM ATE VIII +|| D DVMANNI New Moon crossing the Sun and Venus Mars @ CE+E 0h	January 13, 2002
3. DVM ATE VI ||+ D Moon to cross Sun &Venus Moon to cross CE+E 0h Moon 28 nights	February 10, 2002
3a. DVM ATE X + D Saturn w/ Aldebaran Moon at CE+E 0h 2 nights SGP	February 14, 2002
4. RIV VIII |+| D ANAGANTIO Venus @ SGP Full Moon 14 nights w/Regulus ready to cross CE+E 12	February 26, 2002
4a. RIV VIIII ||+ D ANAGANTIO Venus @ SGP Moon 16 nights at CE+E 12 NGP	February 27, 2002
5. ANAG XII +|| D Moon 20 nights w/ Antares ready to cross GE+E 0	April 1, 2002
5a. ANAG XIII |+| D Moon at GE+E 0	April 2, 2002
5b. ANAG XIII ||+ D Mars close to Pleiades Moon passed GE+E 0 w/ Sagittarius	April 3, 2002
6. ANAG ATE X ++ D Moon w/Venus Mars w/Pleiades Saturn in Horns of Bull	April 14, 2002
6a. ANAG ATE XI || + Moon w/Mars & Pleiades Venus below Saturn in Horns of Bull	April 15, 2002
CVTIOS none	
OGRO None	
7. GIAM ATE IIII +|| D Moon to pass Horns of Bull & GE+E 180 Venus w/ Regulus Mars w/ Jupiter hidden by Sun	July 6, 2002 eclipse full 6.24

7a. GIAM ATE V |+| D AMB Moon in Horns of Bull 27 nights to pass Saturn	July 7, 2002
7b. GIAM ATE VI ||+ D Moon @ GE+E 180 28 nights Venus w/ Regulus	July 8, 2002
8. GIAM ATE X +|| D Moon w/ Venus & Regulus to cross the CE+E 12	July 12, 2002
8a. GIAM ATE XI N INIS R Moon crosses Venus	July 13, 2002
8b. GIAM ATE XII||+ Moon @ CE+E 12	July 14, 2002
9. SIMIV ATE VI |+| D EQVI Venus @ CE+E 12 Saturn close to GE+E 180	Aug. 6, 2002
9a. SIMIV ATE VII||+ D EQVI AMB Venus moving off CE+E 12	Aug. 7, 2002
10. EQVOS ATE VI |+| MD SIMISI Moon w/ Mars & Regulus hidden by Sun Moon 28 nights to cross CE+E 12	Sept. 5, 2002
10a. EQVOS ATE VII |+| D ELEMB AMB New Moon to cross CE+E 12	Sept. 6, 2002
10b. EQVOS ATE VIII ||+ Moon 1 night old @ CE+E 12	Sept. 7, 2002
11. EQVOS ATE XI +|| D AMB Moon 5 nights just passed Venus moving to cross Antares	Sept. 10, 2002
11a. EQVOS ATE XII |+| D Saturn @ GE+E 180 Moon 6 nights moving to Antares	Sept. 11, 2002
11b EQVOS ATE XIII ||+ D Moon 7 nights w/ Antares close to GE+E 0	Sept. 12, 2002
12. ELEMB ATE XII +|| D Moon 6 nights @ GE + E 0 Mercury & Mars crossing CE + E 12 Saturn @ GE + E 180	Oct. 11, 2002
12a. ELEMB ATE XIII |+| D AMB Mercury & Mars @ CE + E 12 Saturn @ GE + E 180 Moon passed GE +E 0	Oct. 12, 2002
12b. ELEMB ATE XIII ||+ D Mars passed CE + E 12 Mercury @ CE +E 12 Saturn @ GE +E 180	Oct. 13, 2002 eclipse full 11.20
EDRIN none	

13. CANTL ATE VIII |+| D CANTLI AMB Saturn @ GE+E 180 Venus w/Mars Moon eclipse new 12.4 just passed GE+E 0 Mercury @ GE+E	Dec. 6, 2002
13a. CANTL ATE XIII ||+ Moon to cross CE+E 0h in Aquarius bag	Dec. 10, 2002
13b. CANTL ATE XIIII |+| Moon @ CE+E Oh Venus w/Mars	Dec. 11, 2002
1. SAMON III +|| D DVM IVO Moon to cross the Horns of the Bull ready to cross the GE+E 180	Year 2 Dec. 14, 2002
2. SAMON VIIII ||+ MD Moon crossed GE+E 180 Sun @ GE+E 0	Dec. 20, 2002
3. SAMON XIII +|| MD Moon to pass CE+E 12	Dec. 24, 2002
4. SAMON ATE II ||+ D TRINVXSAMO Moon to pass Mars & Venus in kite of Libra	Dec. 28, 2002
5. SAMON ATE IIII +|| MD Moon passed Venus & Mars with Antares to cross GE+E 0	Dec. 30, 2002
5a. SAMON ATE V |+| D AMB Moon to cross GE+E 0 28 nights	Dec. 31, 2002
5b. SAMON ATE VI ||+ MD Moon @ GE+E 0 29 nights	Jan. 1, 2003
6. SAMON ATE X +|| MD Moon to cross SGP moving to CE+E Oh	Jan. 5, 2003
6a. SAMON ATE XI |+| MD AMB IVOS Moon at Aquarius before SGP	Jan. 6, 2003
6b. SAMON ATE XII ||+ MD IVOS Moon @ SGP to cross CE+E Oh	Jan. 7, 2003
7. DVM VIII + MD SAMONI Moon w/ Jupiter & Cancer Mars & Venus w/ Antares	Jan. 18, 2003
8. DVM ATE III |+| D AMB Moon passing Venus close to GE+E 0	Jan. 28, 2003
8a. DVM ATE IIII + D Moon & Venus equal distance both sides of GE+E 0 Mars w/ Antares	Jan. 29, 2003
9. DVM ATE VI + D New Moon to cross Sun halfway between GE+E 0 & CE+E Oh Capricorn	Jan. 31, 2003
10. DVM ATE VIII ||+ D Moon to cross CE+E Oh same time Venus to cross GE+E 0	Feb. 2, 2003

10a. DVM ATE X + D Moon @ CE+E Oh Venus @ GE+E 0 Sun in the middle	Feb. 4, 2003
11. RIV ATE X ||+ MD PETIVX RIVRI Sun @ SGP Mars @ GE+E 0 Moon just crossed CE+E Oh Jupiter w/ Cancer	Mar. 5, 2003
11a. RIV ATE XI +| D AMB IVOS Mars @ GE+E 0 Jupiter w/ Cancer	Mar. 6, 2003
11b. RIV ATE XII |+ MD IVOS Moon to cross Pleiades	Mar. 7, 2003
11c. RIV ATE XIII ||+ MD IVOS Moon crossing Pleiades Mars across GE+E 0	Mar. 8, 2003
12. ANAG ATE III +|| D AMB Moon 26 nights to pass Venus Moon halfway between GE+E 0 & CE+E Oh equal distance between Venus/Mars	Mar. 28, 2003
12a. ANAG ATE IIII ||+ D Moon passing Venus	Mar. 29, 2003
12b. ANAG ATE V + D AMB Moon 28 nights passed Venus @ SGP Jupiter @ Cancer	Mar. 30, 2003
13. ANAG ATE XI |+| D AMB Moon 4 nights w/ Pleiades	April 5, 2003
OGRONIOS none	eclipse new 5.31
CVTIOS none known	
14. GIAM ATE IIII +|| D Moon between Pleiades & Aldebaran ready to pass Venus hidden by Sun Sun w/ Saturn just passed GE+E 180	June 26, 2003
14a. GIAM ATE V |+| D AMB Moon w/ Venus & Mercury end of Horns of the Bull	June 27, 2003
14b. GIAM ATE VI ||+ D Moon w/ Mercury @ GE+E 180 followed by two INI marked nights	June 28, 2003
15. GIAM ATE X +|| D Venus just before GE+E Saturn & Mercury other side of GE+E 180 note all this is hidden by the Sun followed by INI mark	July 1, 2003
15a. GIAM ATE XII ||+ D Venus @ GE+E	July 3, 2003
16. SIMIV ATE IIII ||+ MD Moon 26 nights @ end of Horns of Bull ready to cross GE+E 180	July 25, 2003

17. SIMIV ATE VI ||+ D EQVI Moon 28 nights ready to cross Venus hidden by the sun	July 27, 2003
18. SIMIV ATE VIII +|| MD Moon 1 nights ready to cross Regulus Jupiter is close and Mercury on Regulus	July 29, 2003
18a. SIMIV ATE VIIII |+| D AMB Moon 2 nights w/ Regulus & Mercury	July 30, 2003
19. EQVOS VI |+| N SIMIV Venus w/ Sun Moon to cross CE+E Oh 2 nights just crossing Mars	Aug. 12, 2003
20. ELEMB ATE III +|| D AMB EDRIN Sun to cross CE+E 12 Moon just past GE+E 180 is w/Saturn	Sept. 19, 2003
21. EDRIN XII |+| MD Moon 18 nights w/ Pleiades	Oct. 13, 2003
21a. EDRIN XIII |+| MD Moon 19 nights between Capella & Aldebaran	Oct. 14, 2003
21b. EDRIN XIIII ||+ MD Moon 20 nights @ GE+E 180	Oct. 15, 2003
22. EDRIN ATE IIII +|| MD Moon 25 nights w/ Regulus ready to cross CE+E 12	Oct. 20, 2003
23. EDRIN ATE VIIII +|| D Moon 1 night to cross Venus Venus is halfway between GE+E 0 & CE+E 12	Oct. 25, 2003
23a. ERIN ATE X |+| MD SIND IVOS Moon w/ Venus halfway between GE+E 0 & CE+E Oh	Oct. 26, 2003
23b. EDRIN ATE XI ||+ D AMB Moon passed Venus now w/ Antares 10.29	Oct. 27, 2003
24. CANTL ATE VI ||+ D 11.9 Moon 27 nights Venus equal distance from Sun	Nov. 21, 2003 eclipse full 11.9
24a. CANTL ATE XIII +|| D Venus just crossed GE+E 0	Nov. 28, 2003 eclipse new 11.24
24b. CANTL ATE XIIII |+| D Venus just across GE+E 0	Nov. 29, 2003
SAMONIOS none	Year 3
1. DVM VIII |+| MD SAMONI Moon 14 nights 1 night past GE+E 180 w/ Saturn	Jan. 6, 2004

2. DVM ATE VI ‖+ D Moon 27 nights @ GE+E 0	Jan. 19, 2004
3. RIVR VIIII +‖ MD Venus to cross CE+E Oh	Feb. 5, 2004
3a. RIVR X ‖+ MD Venus @ CE+E Oh Moon w/ Regulus	Feb. 6, 2004
4. RIVR ATE V +‖ D AMB IVO CE+E due East/West 8:40 PM/AM	Feb. 16, 2004
4a. RIVR ATE VI |+| MD this is also the time that GE+E 180 reaches zenith due South	Feb. 17, 2004
4b. RIVR ATE VII ‖+ D AMB same as above	Feb. 18, 2004
5. RIVR ATE X IVRIDRIVRI ‖+ M Moon 2 nights old to cross CE+E Oh	Feb. 21, 2004
6. ANAG VIIII +‖ D Moon to cross CE+E 12 just past Jupiter	Mar. 6, 2004
7. OGRO VIIII +‖ MD Venus w/ Pleiades Full Moon just past CE+E 12	April 4, 2004
7a. OGRO X |+| MD Venus between Pleiades & Aldebaran	April 5, 2004
7b. OGRO XI ‖+ D AMB Venus above Pleiades	April 6, 2004
8. OGRO ATE III ‖+ D CVTIO AMB Venus passing beside Horns of Bull Mars in Horns of Bull Moon past GE+E 180 over third qtr.	April 13, 2004
9. OGRO ATE VI +‖ MD Moon crossing CE+E Oh	April 16, 2004
9a. OGRO ATE VII |+| D AMB Moon passing CE+E	April 17, 2004
10. OGRO ATE VIII |+| Venus is outside ecliptic ready to pass Mars in Horns of Bull New Moon	April 18, 2004 eclipse new 4.19
10a. OGRO ATE XIII |+| D AMB Moon w/ Mars & Venus Mars in Horns of the Bull	April 23. 2004
10b. OGRO ATE X IIII ‖+ M D Moon passes GE+E 180 Moon w/ Saturn	April 24, 2004 eclipse full 5.4
11. CVTIO ATE VII +‖ D AMB Moon to pass Pleiades hidden by light of sun	May 17, 2004

11a. CVTIO ATE VIII |+| New Moon w/ Pleiades & Sun	May 18, 2004
11b. CVTIO ATE VIIII ||+ Moon 1 night passed Pleiades hidden by Sun	May 19, 2004 eclipse new 5.19
12. CVTIO ATE XIII +|| D AMB IVO Sun passing Pleiades	May 23, 2004
13. OGMIOS XII ||+ D CANTLI Venus meets w/ Sun end of Horns of the Bull	June 6, 2004
13a. OGMIOS XIII +|| MD SAMONI Venus w/ the Sun	June 7, 2004
14. OGMIOS ATE II +|| MD QVTI IN OGRO Moon just past CE+E Oh to pass Pleiades	June 11, 2004
14a. OGMIOS ATE VI ||+ D SIMIVISONN Moon just past Pleiades Moon between Pleiades & Venus hidden by light of the Sun	June 15, 2004
15. OGMIOS ATE X +|| MD SAMON Sun crossing GE+E 180	June 19, 2004
15a. OGMIOS ATE XII ||+ MD RIVRI Sun just past GE+E 180	June 21, 2004
16. OGMIOS ATE XIIII ||+ D OGRONV Venus w/ Aldebaran & Pleiades	June 23, 2004
17. GIAM VII +|| MD SIMIVI TIOCBR Full Moon has just past GE+E 0 Galactic Center Antares	July 1, 2004
17a. GIAM ATE VI ||+ D Moon 28 nights just past GE+E 180*	July 15, 2004
18. SIMIV III |+| D EQVI Moon passing GE+E 0 GC & Antares	July 26, 2004
19. SIMIV XI +|| D AMB Moon passing CE+E Oh	Aug. 3, 2004
19a. SIMIV XII |+| MD Moon past CE+E Oh Venus close to GE+E 180	Aug. 4, 2004
EQVOS none	
ELEMBIVIOS none	
20. EDRIN XII +|| MD Moon just past GE+E 180 Venus just past CE+E 12	Nov. 1, 2004
20a. EDRIN XIII |+| MD Moon w/ Saturn Venus w/ Jupiter	Nov. 2, 2004
20b. EDRIN XIIII ||+ MD Moon w/ Saturn Venus w/ Jupiter	Nov. 3, 2004

21. EDRIN ATE IIII |+| MD Moon to cross CE+E 12	Nov. 8, 2004
21a. EDRIN ATE V |+| D AMB Moon crosses CE+E 12 to meet w/ Venus & Jupiter	Nov. 9, 2004
21b. EDRIN ATE VI ||+ MD Moon crossed Venus & Jupiter and meets w/ Mars Moon 28 nights	Nov. 10, 2004
22. EDRIN ATE X |+| MD Moon on GE+E 0 CANTLOS none	Nov. 14, 2004
1. SAMON XIII +|| MD Moon w/ Regulus to cross CE+E 12 Moon 20 nights	Year 4 Dec. 31, 2004
1a. SAMON XIIII ||+ MD Moon crossing CE+E 12	Jan. 1, 2005
1b. SAMON XV ||+ Moon crossed CE+E 12	Jan. 2, 2005
2. SAMON ATE IIII +|| MD Moon to cross GE+E 0	Jan. 6, 2005
2a. SAMON ATE V |+| D AMB Moon crossing Antares w/ Mars	Jan. 7, 2005
2b. SAMON ATE VI ||+ MD Moon 28 nights meets GE+E 0 at the same time as Venus & Mercury	Jan. 8, 2005
3. SAMON ATE X +|| MD Moon 3 nights to cross CE+E Oh	Jan. 12, 2005
3a. SAMON ATE X I |+| D AMB Moon to cross CE+E Oh	Jan. 13, 2005
3b. SAMON ATE XII ||+ MD Moon @ CE+E Oh SGP	Jan. 14, 2005
4. DVM ATE X + D Moon 3 nights @ CE+E Oh Mars @ GE+E 0	Feb. 11, 2005
5. ANAG ATE VII |+| D AMB OGRON New Moon meets w/ Venus w/ Sun	April 8, 2005 eclipse new 4.8
5a. ANAG ATE VIII |+| MD QVTI OGRON Moon to meet w/ Pleiades Mars on CE+E Oh Venus w/ Sun	April 9, 2005 eclipse full 4.24
6. OGRON ATE I +|| MD QVTIO Moon halfway between GE+E 0 & CE+E Oh Capricorn	May 1, 2005
6a. OGRON ATE II |+| MD QVTIO Moon past Mars	May 2, 2005

6b. OGRON ATE III ‖+ D AMB QVTIO Moon passed halfway between GE+E & CE+E now close to CE+E Oh Venus close to Pleiades hidden by light of Sun	May 3, 2005
7. OGRON ATE VII |+| AMB QVTIO New Moon, Venus w/ Pleiades hidden by light of the Sun	May 7, 2005
7a. OGRON ATE VIII ‖+ MD OGRO QVTI Moon 1 night passing Venus w/ Pleiades hidden by light of Sun	May 8, 2005
7b. OGRON ATE VIIII ‖+ D AMB QVTIO Moon to cross GE+E 180	May 9, 2005
CVTIOS none	
8. GIAMON ATE XII ‖+ D Moon to cross CE+E Oh	July 11, 2005
9. SIMIV III |+| D EQVI Moon w/ Antares close to GE+E 0	July 17, 2005
10. EQVOS IIII ‖+ D Venus @ CE+E 12	Aug. 16, 2005
11. EQVOS VIIII ‖+ D Moon @ CE+E Oh	Aug. 21, 2005
11a. EQVOS C |+| D Moon just past CE+E Oh	Aug. 22, 2005
12. ELEMB ATE III +‖ D AMB EDRIN Moon w/ Saturn & Cancer Mars w/ Pleiades	Sept. 27, 2005
13. EDRIN ATE IIII +‖ MD Moon 25 nights to cross CE+E 12	Oct. 27, 2005
13a. EDRIN ATE V ‖+ D AMB Moon crossing CE+E 12	Oct. 28, 2005
13b. EDRIN ATE VI ‖+ D AMB Moon @ CE+E 12	Oct. 29, 2005
14. EDRIN ATE VIIII +‖ D AMB New Moon to cross Antares Venus to cross GE+E 0	Nov. 1, 2005
14a. EDRIN ATE X |+| MD SINDIV IVO Venus @ GE+E 0	Nov. 2, 2005
14b. EDRIN ATE XI ‖+ D AMB Venus crossing GE+E 0	Nov. 3, 2005
15. CANTL X +‖ D Moon @ end of Horn's of Bull ready to cross GE+E 180	Nov. 17, 2005
15a. CANTL XI |+| D AMB Moon crosses GE+E 180	Nov. 18, 2005

1. SAMON ATE IIII +‖ MD Moon to cross Antares	Year 5
1a. SAMON ATE V |+| D AMB Moon to cross Antares	Dec. 26, 2005
1b. SAMON ATE VI ‖+ MD Moon on Antares	Dec. 27, 2005
2. DVM VIII [+] MD SAMONI Venus passes the Sun	Jan. 13, 2006
RIVROS none known	
3. ANAG ATE III +‖ D AMB Moon just past GE+E 0 ready to pass Venus	Mar. 23, 2006
3a. ANAG ATE IIII |+| D Moon ready to pass Venus	Mar. 24, 2006
3b. ANAG ATE V ‖+ D AMB Moon w/ Venus	Mar. 25. 2006
4. ANAG ATE X + D Moon passes Sun close to CE+E 0h	Mar. 30, 2006
4a. ANAG ATE XI ‖+ D AMB Moon to pass Pleiades equal distance from Sun as Venus other side Mars @ end of Horns of Bull	Mar. 31, 2006 eclipse full 4.13
5. OGRON ATE XIII |+| D AMB Venus @ CE+E 0h	May 1, 2006
5a. OGRON ATE XIIII ‖+ MD Venus passing CE+E 0h	May 2, 2006
CVTIOS none known	
6. GIAMON ATE IIII +‖ D Sun @ GE+E 180 Moon to pass Venus w/ Pleiades Mars w/ Saturn @ Cancer	June 21, 2006
6a. GIAMON ATE V |+| D AMB Moon w/ Venus & Pleiades Moon very close to Pleiades	June 22, 2006
6b. GIAMON ATE VI ‖+ D Moon passed Venus & Pleiades	June 23, 2006
7. SIMIV ATE IIII ‖+ MD Moon to pass Venus @ GE+E 180	July 20, 2006
7a. SIMIV ATE VI ‖+ D EQVI Moon w/ Venus just past GE+E 180	July 22, 2006
8. SIMIV ATE VIII +‖ MD Moon to pass Saturn halfway between GE+E 180 & CE+E 12	July 24, 2006

8a. SIMIV ATE VIIII |+| D AMB Moon w/ Saturn halfway between GE+E 180 & CE+E 12 hidden by light of the Sun Mars w/ Regulus	July 25, 2006
9. EQVOS IIII ||+ D Sun passing Saturn halfway between GE+E 180 & CE+E 12	Aug. 4, 2006
9a. EQVOS VIIII |+| D Sun just past Saturn halfway between GE+E 180 & CE+E 12	Aug. 9, 2006
9b. EQVOS X ||+ D Sun past Saturn halfway between GE+E 180 & CE+E 12	Aug. 10, 2006
10. EQVOS ATE V +|| D AMB Moon to pass Venus halfway between GE+E 180 & CE+E 12	Aug. 20, 2006
10a. EQVOS ATE VI ||+ MD SIMIVIS Moon w/ Venus halfway between GE+E 180 & CE+E 12	Aug. 21, 2006
10b. EQVOS ATE VII ||+ D AMB Moon passed Venus halfway between GE+E & CE+E 12 close to Cancer	Aug. 22, 2006 eclipse new 8.23 eclipse full 9.7
11. ELEMB XI ||+ D AMB Moon to pass the Pleiades Mars @ CE+E 12	Sept. 10, 2006
12. ELEMB ATE VI +|| D Almost new Moon passes Venus hidden by light of Sun while Sun passes CE+E 12	Sept. 20, 2006
12a. ELEMB ATE VII |+| D AMB Moon meets the Sun @ CE+E 12 Mars & Venus equal distance from Sun both sides hidden by light of Sun	Sept. 21, 2006 eclipse new 9.22
13. ELEMB ATE XII +|| D Moon ready to pass Antares	Sept. 26, 2006
13a. ELEMB ATE XIII |+| D AMB Moon w/ Antares	Sept. 27, 2006
13b. ELEMB ATE XIIII ||+ D Moon passed Antares ready to cross GE+E 0	Sept. 28, 2006
14. EDRIN ATE VI ||+ MD Moon @ CE+E 12 ready to pass Venus w/ Sun & Mars	Oct. 19, 2006
15. CANTL V ||+ D AMB Moon just passed CE+E Oh	Nov. 2, 2006
16. CANTL ATE XIII +|| D AMB IVO Venus w/ Antares	Nov. 25, 2006

1. OGMIOS ATE VI ||+ MD RIVRI Moon passing Antares where Jupiter & Mars are also	Year 6 is year 1 Dec. 17, 2006
2. SAMO III +|| D EXINGIDVM IVOS Moon to pass the Pleiades	Dec. 29, 2006
3. SAMO ATE VI ||+ M D Moon 27 nights crossing GE+E 0 where Mars is. Venus halfway between GE+E 0 & CE+E	Jan. 16, 2007
3a. SAMO ATE VIII +|| D DVMANNI AMB New Moon passes the Sun	Jan. 18, 2007
4. DVM ATE VI ||+ D Venus passes SGP just before CE+E Oh Moon passes Mars	Feb. 14, 2007
5. DVM ATE X + D Moon @ CE+E Oh w/ Venus Jupiter w/ Antares	Feb. 19, 2007
6. RIVR VII |+| D ANAGANTIO Full Moon total eclipse passing NGP Venus above the Sun PM	Mar. 3, 2007 eclipse full 3.3
6a. RIVR VIII ||+ D ANAGANTIO Moon @ CE+E 12 Sun to pass SGP	Mar. 4, 2007 eclipse new 3.19
7. ANAG XII +|| D Moon 19 nights w/ Antares Venus w/ Pleiades	April 6, 2007
7b. ANAG XIIII ||+ D Moon @ GE+E 0 Venus w/ Pleiades	April 8, 2007
8. ANAG ATE X ++ D Moon 3 nights w/ Venus & Pleiades Saturn close to zenith and Regulus	April 19, 2007
8a. ANAG ATE XI ||+ D Moon past Pleiades & Venus, is end of Horns of the Bull ready to cross GE+E 180	April 20, 2007
CVTIOS none	
9. GIAM ATE IIII +|| D Moon w/ Horns of the Bull ready to cross GE+E 180 Mercury is @ GE+E 180 Venus & Saturn w/ Regulus	July 11, 2007
9a. GIAM ATE V |+| D AMB Moon 28 nights w/ Mercury@ GE+E 180	July 12, 2007
9b. GIAM ATE VI ||+ D New Moon passed GE+E 180 w/Sun	July 13, 2007

10. GIAM ATE X +|| D Moon 3 nights just passed Saturn, Venus, Regulus ready to cross CE+E 12	July 17, 2007
10a. GIAM ATE XII ||+ D Moon 5 nights passed CE+E 12	July 19, 2007
11. SIMIV ATE VI |+| D EQVI Moon 29 nights close to Sun halfway between GE+E 180 & CE+E 12 Moon equal distance from Sun as Venus & Saturn hidden by Sun	Aug. 11, 2007
11a. SIMIV ATE VII ||+ D EQVI AMB new Moon passing Sun, Saturn, Venus	Aug. 12, 2007
12. EQVOS ATE VI +|| MD SIMISI new Moon to cross CE+E 12 hidden by Sun	Sept. 10, 2007 eclipse new 9.11
12a, EQVOS ATE VII |+| D ELEMB AMB Moon crossing CE+E 12 hidden by the Sun	Sept. 11, 2007
12b. EQVOS ATE VIII ||+ D ELEMB Moon passed CE+E hidden by Sun	Sept. 12. 2007
13. EQVOS ATE XII +|| D AMB Moon to cross Antares Sun passing NGP	Sept. 15, 2007
13a. EQVOS ATE XIII |+| D Moon moving closer to Antares Sun below NGP & Coma Berenices	Sept. 16, 2007
13b. EQVOS ATE XIIII ||+ D AMB Moon w/ Antares & Jupiter Sun passed NGP	Sept. 17, 2007
14. ELEMB ATE XII +|| D Moon @ GE+E 0 Venus w/Saturn	Oct. 16, 2007
14a. ELEMB ATE XIII |+| D AMB Moon passing GE+E 0 Jupiter close Venus passed Saturn	Oct. 17, 2007
14b. ELEMB ATE XIIII ||+ D Moon passed GE+E 0	Oct. 18, 2007
EDRIN none	
15. CANTL ATE VIII |+| Jupiter @ GE+E 0 Saturn w/Coma Berenices	Dec. 11, 2007
16. CANTL ATE XIII ||+ D AMB IVOS Mercury to cross the Sun Jupiter @ GE+E 0 Moon to cross CE+E Oh	Dec. 15, 2007
16a. CANTL ATE XIIII |+| D IVO DIB CANT Mercury w/the Sun Moon	Dec. 16, 2007

NOTES FOR CHART

The T marks w/ GE+E & CE+E
GE+ E is the intersection of the Galactic Equator and the Ecliptic
CE+E is the intersection of the Celestial Equator and the Ecliptic

Appendix III

Of K'uk'ulcán and Quetzalcóatl

Venus in Mesoamerica

R. P. Hale

In the Western Hemisphere, the varied Mesoamerican cultures of what is now Mexico viewed and evolved their concepts associated with Venus quite differently from cultures in Africa, Europe and Asia. No one article could possibly explore all of the aspects of the Mesoamerican Venus, and this essay won't attempt it. While there are some parallel concepts, the vast majority of Mesoamerican views concerning the third brightest celestial object seen from Earth are completely different, and in many cases opposite.

We can attribute most of this evolutionary separation to geography. When the last great Ice Age was ending, rising water levels from the melting ice flooded the Bering pathway that had connected Siberia and Alaska, leading to an effective isolation of the two American continents that would be relieved only by seafarers. Some of these documented their accidental visits to these unknown lands, including the expeditions of the Chinese admiral Zhèng Hé (Cheng-Ho) in the early 1400s, the Viking voyages to Greenland and Labrador, and most familiarly, the voyages of Cristobál Colón—Columbus, that led the disastrous European invasions that overwhelmed and ended Native cultural dominance in the Americas.

It can be difficult to get a firm handle on the myriad Mesoamerican Venus concepts. One reason is that an enormous number of cultures evolved and lived (and still do today) in relative close proximity to each other in a surprisingly small area, roughly extending from the northern deserts of the Tropic of Cancer and Teotihuacán to the Yucatán peninsula and Central regions almost to Panama. There were some thirty groups within the Maya alone. They interacted in the usual manner, through language, trade, conquest and sporadic travel, exchanging their stories and beliefs in the process. Successive or conquering tribes often added the gods and customs of their predecessors to their own lore. Such would lead to common stories with many cultural/ regional variations, which we see in the widespread—and often very confusing—accounts referring to the visible planet Venus in various common-rooted incarnations, the most well known being the major deity K'ulk'ulcán, who would later become Quetzalcóatl.

Another reason for all the overlapping confusion would derive from the European conquests themselves. The Spanish *conquistadores* came *por Dio y oro,* for God and Gold (gold was more important), and built a record of appropriation and genocide on two continents. Europeans in general saw themselves as supplanting the primitive peoples, who *deserved* to be pushed aside by the more advanced and civilized whites, as evidenced in the 19th-century "doctrine" of "manifest destiny." Missionaries overall had no tolerance for the old beliefs, condemning them as false and against the new God, often invoking the secular authority to punish religious malefactors. Mesoamerican peoples had themselves used the same tactics in war, eliminating the priests and ruling class upon subduing a rival. But the overall result of European conquest was massive cultural loss, manifest in the coordinated Spanish mass burning of Mexica, Maya and Mixtec codices that had been collected over the centuries. The old lore survived in furtively recounted stories and fragments, sometimes documented at the behest of friars such as Bernardino de Sahagún who was very interested in the old beliefs. One major Maya account that survived in this manner is the *Popol Vuh.* And the bishop Diego de Landa, who oversaw the mass destruction of Yucatec Maya books, would happen be the one who would leave modern epigraphers a major key to deciphering the ancient Maya glyphs.

The Mesoamerican cultures developed the famous calendar system, brought to its height by the time-obsessed Maya, whose system of interlocking time-wheels was very accurate for that period. Looking back from our own time and place, we tend to forget that their superb celestial and time-keeping feats were based on divination and astrology, not on the scientific principles of today's astronomy.

Venus and the Sun were major factors in these calendar systems. Evidence points to the Olmec and Zapotec cultures as the one who built the basis of what the Maya groups would evolve into their surprising assembly of interlocking calendar wheels. Mesoamerica was the only place on Earth where cultures counted in base-20, the *vigesimal system*, and where the concept of zero predated the Hindu mathematicians. This number system allowed for large-scale calculation but had one major shortcoming: there was no concept of fractions, so all calculations and concepts could be based solely on whole numbers—you either had a whole ear of corn or a whole cocoa bean, or you didn't! But their unique concept of zero was twofold: the familiar state "of nothingness" combined with the Yucatec Maya term *bix-*

baal, literally "in progress," meaning that you are moving from nothing to something. Think of embarking on a trip: you are not at your destination until you arrive there.

Like so many world cultures, the Sun—*Ah K'in*—would provide the major yardstick for Mesoamerican calendars. Venus, rather than the Moon, was the second most important celestial body to Mesoamericans, who incorporated its motions into the calendars. These observed motions of Venus, called *Noh Ek'* (Great Star), and *Xux Ek'* (Wasp Star) and many other names, determined the propitious time for major undertakings, particularly war. The Maya were well aware of the planet's synodic period, averaging 584 Earth days, and the *Dresden Codex* almanac, one of the few surviving Maya books, devotes six full pages to Venus observations.

Unlike the ancient Greeks who belatedly realized that the "morning" and "evening star" were really the same body, the Mesoamericans knew it was a single celestial object that passed through a cycle of four distinct phases, even though they did so without ever truly understanding what caused these phases. The *Dresden Codex* tables list Venus as a Morningstar for 236 days, followed by a 90-day period of invisibility when the Maya believed it was in the Xibalba underworld. Then came a 250-day period as the Eveningstar, followed by another 8-day period when it again sank into the underworld. Even though the times for these individual phases are definitely not as precise as indicated above, the complete Venus cycle according to the Maya totaled 584 days, remarkably close to the planet's 583.92 day synodic period recognized by modern astronomers. This figure is a mean value, however, because the synodic period varies from 579.6 days to 588.1 days within a given five-year period. Later Maya observations would lead to their being able to calculate this period to within a hundredth part of a solar day, using whole numbers to find repeating cycles over very extended periods of time.

Mesoamericans—and Maya—didn't have our understanding of the solar system with the planets orbiting the Sun. Venus is the second planet out from the Sun and we view it from the third planet out, so Venus (and Mercury) will never "stray" far from the Sun as seen from Earth. The Morningstar phase represents Venus moving from our side of the Sun to the opposite side of its orbit (superior conjunction), so its disappearance in the solar glare lasts for quite a while. When we next see it emerge, it accompanies the setting Sun and we see it as the Eveningstar as it moves away from and then back toward the Sun. The second disappearance (inferior conjunction) from the sky is shorter because the faster-moving Venus is overtaking and passing the Earth while both planets are on the same side of the Sun. This short period of invisibility ends with Venus' reappearance as Morningstar. Remember too that both Venus and Earth move in their respective orbits around the Sun rather like racecars in their own separate tracks, and that Earth's 23.5° axial tilt further complicates our view of the motions of Venus relative to the horizon.

In their observations and records, the Maya early realized that five Venus periods (5 x 584 days = 2920) equaled eight Earth solar years (8 x 365 days = 2920: note the absence of the leap year). So, at the end of every eight of their "Vague Years"—the name given to the 360+5-day period used throughout Mesoamerica—they could anticipate Venus' appearance in the same place in the sky that it had eight years earlier.

Many find it surprising that Mesoamericans would seemingly exalt Venus at the expense of the Moon. In fact, the Mexica ("Aztec") held that the Moon deity began as a braggart god who was really a coward: he was afraid to leap into the sacrificial fire at the start of the current Fifth Era. Had he done so, Tecuciztecatl would have become the Sun, but after several failed attempts, he was shoved aside by the weak, ill god Nanahuátl, who leapt quickly into the sacred fire and immediately vaulted skyward as the Sun. Humiliated, Tecuciztecatl jumped in right after, to rise as a second Sun. Quetzalcóatl— Venus—wasn't having any of *that,* and he flung a rabbit into Tecuciztecatl's face to dim it into that of the Moon, who would also suffer cyclical growth and death. Quetzalcóatl would subsequently sacrifice all of the other gods gathered at the sacred fire in order that their collective strength would feed Nanahuátl the new Fifth Sun, who took the name Tonátiuh.

Such "downplaying" of the Moon and its cycles could have some basis in the number system and its lack of fractions, as whole numbers do not accommodate the ~29.5-day lunar cycle, which itself varies in period and does not coordinate to the solar year, which also includes a fraction. It wasn't until the seventh century that some of the Maya groups, using centuries' worth of observations, would count whole days until lunar cycles and whole numbers added up. In doing so, they came up with the most accurate value for the lunar cycle before the atomic clock, with 149 lunations taking up 4,400 days. Dividing the latter by the former yields a value that is 0.0004 from the current atomic-clock value—noting again, that the Maya could not have done such division to obtain that value.

The Maya found other numerological possibilities. Divide a Venus synodic period by 8 to get an interval of 73 days, which is the same result if you divide a 360+5-day "Vague Year" by five. Adding 8 and 5 gives the sacred number 13, and multiplying that by the other sacred number 20 gives 260, the number of days in the sacred *Tzolkin* calendar, and 73 *Tzolkin* periods equal 18,980 days, or 52 "vague years" which was also the period of "the binding of the years" ritualized by the Náhua peoples centuries later. Meshing the 18,980-day period with the 584-day cycle of Venus yields 104 "Vague Years." While 8 does not evenly divide into 52, it does with 104, so 146 *Tzolkin* equals 104 "Vague Years"—and 65 Venus cycles as well.

However, even though finding the synodic period was an early astronomical accomplishment, the cumulative details of the movements of Venus would not be fully figured out until the seventh century CE, as noted in the tenth-century *Dresden Codex.*

THE GOD OF THE MAYA

To the Mesoamericans overall, the planet Venus variously represented a powerful male deity that incorporated creation, fertility, time cycles, regeneration, resurrection, sacrifice—and war. Its celestial movements led the Maya to see Venus as the companion of the Sun, and to the southern Maya, Venus was more important than the Sun itself. The strikingly white light of Venus had nothing to do with love, maternity or the Female Being. To Mesoamericans, the color white represented death and the direction North with its fearsome deserts. North was also where the savage barbarian *Chichimeca* tribes lived, one raiding band of whom would in time become the Mexica (Aztec). The Maya considered the heliacal rising—first visibility—of the brilliant "star" in the pre-dawn sky as a very evil sign, and kings would plan wars upon seeing this, and many commemorations of these "star wars" are carved into stelae from Palenque, Naranjo and Tikal to Copán and the surrounding areas.

The main temple at the early Maya site of Cerros, in Belize, depicts the earliest known Classical Maya theological version of Venus as the companion to the Sun. Modern scholars have come to realize that Yax-Balam, the Sun Jaguar, represented much more than just the celestial body itself. Yax-Balam is the younger of the Ancestral Hero Twins, the sons of the double-headed Great Cosmic Monster, and his older brother Hun-Ahau was manifested in the planet Venus who attends the Sun as Morningstar and Eveningstar. The façade carvings on the main temple show the setting Sun Jaguar with Eveningstar overhead to the left of the stairway, and the right side shows Morningstar over the rising Sun. A painted bowl elaborates the early Maya view of Venus, from his inception with his Sun Jaguar twin within the Great Cosmic Monster to his rise as Eveningstar Chac-Xib-Chac following his long disappearance (death) in the Xibalba Underworld. And from Chac-Xib-Chac's head sprouts the *Wacab Chan,* the World Tree.

Venus had several glyphic Maya symbols that would remain unchanged for the most part within Maya groups and over their history. This glyph was paired with the Sky symbol and was always present in the ceremonial *Sky Bar* regalia cradled by Maya kings as well as on royal headdresses, pectorals and breechclout coverings, a powerful representation of the royal mandate from Heaven. The paired glyphs also appear on the sarcophagus lid of the 7th-century king K'inich Jinaab' Pacal of Palenque.

To the first century CE northern/Yucatec Maya, the planet Venus would become the visible manifestation of the Feathered Serpent, Ku'k'ulcan or Q'uq'umatz, who apparently evolved from the earlier Vision Serpent. Throughout Mesoamerican iconography, serpents represented the important connection between gods and humans and the Maya used this image in carvings to depict the communion between their rulers and the gods. Maya rulers regularly and often bled themselves in ritual, dripping their blood onto strips of paper that they then burned in sacred fires to observe the serpentine pattern of the curling smoke. In their trances, they saw the smoke as the Vision Serpent that directly communicated messages from the gods that the ruler then imparted to the people, or, portents of how an upcoming war or event would turn out.

THE GOD OF THE MEXICA: STORIES INTERTWINED WITH STORIES

In Náhuatl Mexica lore, Venus had two representations, that of Quetzalcóatl Tlahuizcalpantecuhtli , Morningstar; and his twin brother/evil alter ego Xólotl, the Eveningstar. Morning and evening star are opposite views of what they knew were the same body, and Morningstar was the good and beneficial manifestation. Xólotl, as the evening star, represented the dark side of things as well as being the god of ill luck—even while he was also charged with guarding the Sun on his nightly trips through the Mictlán Underworld. As with most known Mesoamerican cultures, the Mexica tended to schedule their wars around the cycle of Venus, fighting when Xólotl Eveningstar was visible and also during conjunctions with the Sun when Venus would be invisible and "dead."

The Mexica used many aspects of the earlier Mesoamerican/Maya calendars, but in reduced form and without the number zero. The Maya's Long Count had been abandoned for over six centuries by now, so the two longest cycles were the 52-year Calendar Round and the 104-year Double Cycle, derived by intermeshing the 52-year cycle with the 584-day period of Venus.

The Náhuatl name *Quetzalcóatl* means *Feathered Serpent* and was roughly equivalent to the old Feathered Serpent deity that arose around the first century CE in the Teoticahuán region. Snakes were long associated with visions, Earth, and kingly power, and *feathered* serpents could fly. The image of a feathered serpent long represented sacred visions and appears in innumerable Maya and Olmec carvings associated with kings. Quetzalcóatl was credited with the creation of humans at the beginning of this current Fifth Sun era, and also represented resurrection, the winds, knowledge, traveling merchants, the arts, and the discovery of maize, which was held from the earliest times as the Mesoamerican staff of life. He was the patron deity of the Mexica priesthood, and his temple was sited in the *Templo Mayor* central complex in México-Tenochtitlán, capital city of the Mexica, facing the mammoth dual pyramid of Tláloc and Huitzilopóchtli.

Quetzalcóatl had perennial importance among the Mexica and related Nahua groups, who saw him as one of the four sons of the dual high god Ometeotl/Ometecuhtli -Omecihuatl , Two-God, He and She of Everything, the unapproachable Highest God of all. He was closely associated with and even exchanged identities with the wind god Éhecatl. He would become Quetzalcóatl-Éhecatl to represent the divine wind/human breath of life, creation, and for the power to withstand the dangers of his frequent trips to the Underworld.

The "good" Quetzalcóatl/Morningstar created modern humankind as part of that of the current Fifth Era. He had to travel to the Underworld of Mictlán to gather up the human bones down there after presenting himself with the request to do so to the Death God Mictlantecuhtlitzin. In an echo of the Maya *Popol Vuh* account, the Death God gave Quetzalcóatl several tasks to complete successfully, which he did. But upon seeing Quetzalcóatl actually gathering up the bones to take to the Upper world, the Death God changed his mind and gave chase, but Quetzalcóatl successfully evaded capture and carried the bones to the surface. He ground them up, wet the bone powder with his own blood, and from this mixture molded humankind.

As mentioned earlier, Quetzalcóatl was present at the Fifth Era creation of the Sun, when all the gods gathered at Teotihuacán to light the Sacred Fire and initiate the new era. They selected the strong, swaggering god Tecuciztecatl to sacrifice himself in the fire to become the new Sun, and he tried repeatedly and failed, for he feared the pain. Then the little ill god Nanahuátl shouldered Tecuciztecatl aside to jump into the Sacred Fire to rise from the flames as the Sun. Humiliated by this, Tecuciztecatl cremated himself too and became a second Sun, but Quetzalcóatl punished him by hurling a rabbit into Tecuciztecatl's face, thus dimming his light to become the lesser Moon. He further decreed that Tecuciztecatl would suffer cyclical death and rebirth as seen in the lunar phases. Nanahuátl the New Sun proved unable to move in the sky, lacking the strength to rise past the dawn, so it fell to Quetzalcóatl to sacrifice all of the other gods present in order to give their collective strength to the new Sun to travel the sky.

This carries through on the earlier Maya belief that Venus was in many respects more important than the Sun itself. Nanahuátl after his resurrection as the Sun still needed the strength to travel the sky, and only Quetzalcóatl could enable him to do so. And Quetzalcóatl-Venus still closely attends and even leads the Sun as Morning Star. The dark Venus Xólotl guides the Sun on his nightly Mictlán underworld journey. And Quetzalcóatl-Venus used the sacrifice of the other gods to establish human sacrifice as necessary to feed the Sun and other deities who needed human hearts and blood to sustain them—for, if the gods could give themselves, could humans do any less? This creator who gave greatly to humankind also took away in equal measure.

Quetzalcóatl's own Death journeys to Mictlán occur when Venus disappears from view during inferior and superior conjunction with the Sun. The inferior conjunction, when Venus and Earth are on the same side of the Sun, has about eight days of invisibility and was called the Little Death. Superior conjunction occurs when Venus on the other side of the Sun from Earth and disappears for a far longer time, and was called the Great Death. The Little Death always follows a Xólotl Eveningstar appearance and thus always heralds the Tlahuizcalpantecuhtli Morningstar apparition. Some Mixtec and later Mexica accounts equate the Little Death with Quetzalcóatl's trip to the Mictlán underworld. Venus' regular reappearance after such invisibility evokes resurrection, and the 13th-century Mixtec *Codex Borgia* depicts Quetzalcóatl as being so powerful that he can dance back-to-back with the Death God Mictlantecúhtlitzin—and live. Here he appears as Quetzalcóatl-Éhecatl, adding the powers of the wind god Éhecatl to his own and representing the breath given humans in order to live.

Venus-Quetzalcóatl appears prominently in the most famous work of Mexica iconography, *In Caxtolcuaúhtli Xipóuhuoálli* , The Eagle Bowl, or, *In Tépetl in Tonatiúhtzin* , the Stone of the Sun, wrongly called "the Aztec calendar." It is no such thing; rather, it represents Mexica cosmography and we read it from the center out. There are many levels of meaning within the layered imagery beyond the visual. In the center is Tonátiuh the Sun with his clawed hands grasping human hearts, surrounded by the symbols of the four previous ages and further superimposed over the glyph representing our current Fifth Era. But underlying all this symbolism is the feather-fringed triangular shape that represents Venus-Quetzalcóatl, whose proximity on the carving shows his association and importance to the Sun. Surrounding Tonátiuh and Quetzalcóatl are circular bands of the count of days, then the starry heavens, symbols of sacrifice and offerings, the Sun's own light rays, and finally two enveloping Xiuhcóatl feathered fire serpents that cradle the Earth on the bottom and indicate the Venus-derived 104-year Double Cycle towards the top.

THE *CODEX CHIMALPOPOCA*: INTERTWINING VENUS AND AN EARTHLY KING

Another Náhua Quetzalcóatl-Venus account comes from the c. 1600 *Codex Chimalpopoca,* which combines the god Quetzalcóatl with stories of a namesake Toltec ruler of a later period. Here, the Suns for the four previous eras were directly influenced by Quetzalcóatl-Venus. He began by literally kicking his brother Tezcatlipoca out as Sun to end the First Era, who repaid Quetzalcóatl by doing the same to him to end the Second. The rain god Tlalocanteúctli became the Third Sun, whose era was destroyed by fire sent by Quetzalcóatl, and the rain god's wife, Chalchiuhtlicue, became the Fourth Sun. This era ended in a flood, "the skies came falling down," with one human couple surviving only to be later turned into *itzquintli* dogs for disobedience. When the floods receded, the heavens were raised back up by Quetzalcóatl and Tezcatlipoca. The latter invented fire, followed by Quetzalcóatl's trip to Mictlán to retrieve the bones to repopulate the world, and that was followed by the emergence of the Fifth Sun at Teotihuacán. The *Codex Chimalpopoca* related that an earthly human massacre accounted for the blood to feed the Sun, rather than Quetzalcóatl's sacrificing the other gods.

Further into the *Codex Chimalpopoca,* the Venus Quetzalcóatl becomes intertwined with actions of the legendary Toltec namesake king. Known also as Cē Ācatl, this king initiated ritualistic human sacrifice during a 160-year rule of the Toltec capital city of Tollán, whose ruins remain in central Mexico, and "he was held to be a god [by Tollán], whom he taught many good things, constructing temples for himself and other monuments, and he lived as a god of that country for 160 years. This Cē Ācatl lived as a lord of Tollán and built a great temple, but four years later. . . the god Tezcatlipoca came to him and advised him that he would have to leave."

Cē Ācatl Quetzalcóatl had to leave because he committed the sin of drunkeness and incest with his sister, and his remorse over his acts led him to abdicate his kingship and leave Tollán. He set out but became ill and died on the seacoast in Tlapallan. Following his cremation, distressing omens convinced the Toltec that their day was done and that the Mexica (Aztec) would inherit their lands.

The *Codes Chimalpopoca* continues (7: 27 – 7: 43):

> . . .when he got to the ocean. . . then he halted and wept and gathered up his attire. And when he was dressed, he set himself on fire and cremated himself. And so the place where Quetzalcóatl was cremated is named Tlatlayan [the place of burning].
> And they say that as he burned, his ashes arose. And what appeared and what they saw were all the precious birds, rising into the sky. . .
> And as soon as his ashes had been consumed, they saw the heart of a quetzal [bird] rising into the sky. And so they knew he had gone into the sky, had entered the sky.
> The old people said he was changed into the star that appears at dawn. Therefore they say it came forth when Quetzalcóatl died, and they called him Lord of the Dawn [Tlahuizcalpantecuhtli].
> . . .So it was after eight days that the morning star came out, which they said was Quetzalcóatl. It was then that he became lord, they said."

This account was related to the Spanish some eighty years after the invasion of Mexico, and the Quetzalcóatl-Venus story had already evolved further with the Mexica culture.

The historic Toltec kings assumed the name Quetzalcóatl as one of their titles, and one of them was the actual ruler Topiltzin Cē Ācatl Quetzalcóatl, and the best evidence places him around the early tenth century. He did rule well and long to become acclaimed as a living god, and he actually *ended* human sacrifice for the Toltec rather than initiating it. Some accounts also have him atoning for drunkenness and incest by abdicating and going to the coast, while others have him traveling across Mesoamerica to establish the cult of the Feathered Serpent.

In the established manner of revision and exculpation, the Mexica revised this Toltec story and combined it with earlier Quetzalcóatl-Venus accounts. They had an almost worshipful respect for the Toltec culture and their kings, which the Mexica had indeed supplanted, and Mexica artists had the title *Tolteca.* But the Mexica's long history with human sacrifice dating back to their Chichimeca nomad days did not allow them to cease that practice, so the stories had to be molded to fit Mexica interests.

Quetzalcóatl-Venus retained an almost unchanged story for the four previous Eras, but the Mexica have him initiating god- and human sacrifice at the start of the current Fifth Era. Confusing the matter further, the Mexica had the earthly Toltec Cē Ācatl Quetzalcóatl sacrificing himself at the end, to have his ashes rise in the form of sacred birds to become the Morningstar Venus. In the process the resurrected Quetzalcóatl-Venus not only resumed human sacrifice but acquired an evil alter ego as the Eveningstar Xólotl. And to the end of the Mexica Empire, incest and public drunkenness were unforgiveable capital crimes. The exception to the latter was that only the aged were permitted to get intoxicated and escape punishment.

FURTHER SPANISH EVOLUTION OF QUETZALCÓATL-VENUS

Much has been made of the story of Quetzalcóatl-Venus setting sail from the coast on a *cóatl* raft of intertwined snakes, promising to return some day in the future, and with the arrival of Hernán Cortéz, this legend was to create all manner of trouble for the emperor Motecuhzoma II Xocoyotzin and the Mexica Empire,.

The problem with this account is that there is practically no evidence that this story predates the Spanish invasion of 1519.

While the Spanish were systematically destroying the cultures of Mesoamerica, they followed a common Church practice of incorporating local stories and lore into the conversion process, which remains today in some Mexican saints being incarnations, in some form, of the earlier Aztec gods, and Mexica ceremonial observances becoming Church holy days. Evidence from the friar Sahagún and other sympathetic monks point to the Spanish allowing the Quetzalcóatl story and other lore to continue under their auspices—with some strategic revisions. And after all, this too had been Mesoamerican conquest practice from the earliest times.

Quetzalcóatl was sometimes depicted in codices as a heavily bearded pale-skinned or albino man. Albinos were very sacred to the Mexica—but slated for sacrifice on account of that sacredness. Mexica men could grow sparse beards and there are some images of warriors with short beards, such as in the pre-invasion *Codex Borgia* and post-invasion *Codex Mendoza*.

Post-invasion accounts including the *Codex Chimalpopoca* could and did mention the Quetzalcóatl "return" legend and its consequences. This version omits Quetzalcóatl's sacrificial-pyre cremation and removes the Venus aspect, replacing those with his weaving snakes together to make a raft and sailing off to the East. Before setting off, the pale-skinned, bearded Quetzalcóatl promised to return some day and take over the rule of Mexico.

The Spanish did sail in from the East, and they were pale-skinned and bearded to the fashion of their time. They also brought horses which thoroughly frightened the natives at first as they had never seen those before, and the mounted and armored conquistador was a formidable figure. They also had arquebuses—guns—which could kill at a distance with the loud explosion and the bullet that traveled too fast to be seen.

There is no doubt that some would see the Spanish as Quetzalcóatl and his followers returned to rule Mexico. But though Motecuhzoma was noted by his own subjects as an extremely nervous and superstitious king, there is no evidence that he or his far cannier priests and advisors would see the Spanish visitors as representing the return of Quetzalcóatl. The emperor had already experienced highly distressing signs and portents, including a great comet, over the past couple years, but while the signs all pointed to catastrophe, none of these signs pointed to Quetzalcóatl, for Venus still appeared completely normal in the sky.

So the Spanish could use an opportunity based on an ancient, deep-rooted legend to legitimize their entry into and rule over the Mexico territory. They could avoid having to use more violence in subjugation simply by invoking their presence as predicted by Quetzalcóatl himself, and get the natives to believe that, which in turn would make them easier to control.

Even today, outside groups have attempted to appropriate Quetzalcóatl to further their own aims and agenda. One religious group invokes the pale and bearded Quetzalcóatl—separated from his Venus aspect—to say that he was actually the resurrected Jesus Christ visiting the New World, clearly ignoring all native accounts that contradict this. Others, of the so-called "New Age" groups in particular, have tried to appropriate K'uk'ulcan-Quetzalcóatl—again, omitting the Venus connections—to support their spurious and fabricated "2012 calendar end," in appalling deliberate ignorance of Maya history, languages and cultures, and especially of the fact that none of the Maya groups had *any* stories about an "end" of any kind.

WORKS CITED

A Forest of Kings, 1990, Linda Schele and David Friedel, *and other works by* Linda Schele, David Friedel, Tatiana Proskouriakoff, David Stuart and David H. Kelley.
Codex Florentine, Bernardino de Sahagún (translation).
Codex Borgia, Codex Nuttall (facsimile publications and online imagery).
Cycles of the Sun, Mysteries of the Moon, 1997, Vincent H. Malmstrom.
Echoes of the Ancient Skies, 1994, Edwin C. Krupp.
Skywatchers, 2001, Anthony F. Aveni.
The Real Venus-Kulkulcan in the Maya Inscriptions and Alignments, 1986, *Sixth Mesa Redonda de Palenque,* A. Aveni.
Astronomy in the Mexican Codex Borgia, 1997, *Archaeoastronomy* No. 24, A. Aveni.
Exploring Ancient Skies: An Encyclopedic Survey of Archaeoastronomy, 2005, David H. Kelley, Eugene F. Milone.
Religions of Mesoamerica, 1990, David Carrasco.
History and Mythology of the Aztecs: The Codex Chimalpopoca, 1992, John Bierhorst.
Aztecs: The People of the Sun, 1958, Alfonso Caso.
World Archeaoastronomy, 1989, J. Sosa; Anthony Aven, ed.
University of Arizona, Department of Anthropology, Tucson, AZ.
Universidad de México *and* Museo de Antropológia e Historia, Ciudad México.
Foundation for the Advancement of Mesoamerican Studies (FAMSI), www.famsi.org/, Los Angeles County Museum of Art.
NASA, www.nasa.gov/, 2012: Beginning of the End or Why the World Won't End?; The Great 2012 Doomsday Scare, 2011.

Plates

Plates

Figure 1. Aphrodite and Eros on Goose, ca. 380 BCE; made at Tarentum, Italy; ht. 18.5 cm (7.25"); British Museum #GR1903.4-12.1.

Figure 2. Ishtar with wings - Akkadian; cylinder seal from the British Museum, BM No. 1891.5-9.2553. ca. 2550 BCE. Courtesy of the British Museum. Photograph by Gregory L. Dexter.

Figure 3. "Durga Dancing," Central India. Gurjara-Pratihara. Tenth century CE. British Museum, Oriental Antiquities. 1872.7-1.82. Courtesy of the British Museum. Photograph by Gregory L. Dexter.

Figure 4. Bronze figure of Hathor with head of a cow. Fourth century BCE. Sera-peion, Memphis, Egypt. ht o,95; lg o,63. Louvre, AF 303. Courtesy of the Musée du Louvre. Photograph by Gregory L. Dexter.

About the Authors

Morgan Llywelyn is the author of thirty nine books which have sold over forty million copies worldwide. She writes historical fiction and historical non-fiction with a genius for making the past come alive. Although her heritage is clearly centered on her Irish roots and cultural identity, much of her work spans many other cultures and time periods. Three of Llywelyn's works, *Druids* (Del Rey 1992), *The Greener Shore: A Novel of the Druids of Hibernia* (Del Rey 2007), and *The Elementals* (Tor Books 2003) include a study of the Druids and/or a vivid depiction of cultural change for many peoples that brings forth characters of great resolve. In her book, *The Elementals*, she identifies the need for new cultures to emerge from the past and predicts future cultures that will emerge from our current global consciousness. She has been more than gracious to write a foreword to this study about creating new mythologies and new worlds complete with the image of the goddess and the balance of the natural world.

Helen Benigni began writing a series of books of which *The Mythology of Venus* is the culminating study. The first book, *The Myth of the Year: Returning to the Origin of the Druid Calendar* (University Press of America, 2003) discusses the mythology of the sun, the moon and the constellations in the ancient Celtic and Greek calendars. The second book, *The Goddess and the Bull: A Study in Minoan-Mycenaean Mythology* (University Press of America, 2007), focuses on the information translated from both calendars on the cycles of the moon and the moon goddess dominant in the ancient cultures of both the Mediterranean and Europe. *The Mythology of Venus* completes a study of a triskele of celestial bodies of the sun, the moon and Venus which reveals the underlying mythology and structure of several ancient calendars. Although her Ph.D. from IUP is in American Literature, her research has been in comparative mythology; she is currently a full professor at Davis and Elkins College in Elkins, West Virginia.

Miriam Robbins Dexter is an editor and researcher *par excellence* who is devoted to the study of the goddesses of ancient cultures. She received her Ph.D. in Indo-European studies from UCLA and her knowledge spans the fields of comparative linguistics, archaeology, and mythology. Her long list of publications includes the two texts, *Whence the Goddesses: A Source Book* (Pergamon Press Athene Series, 1990) and *Sacred Display: Divine and Magical Figures of Eurasia* (Cambria Press, 2010). As editor of the Proceedings of the Annual UCLA Indo-European Conference for seven years and author of more than thirty scholarly articles in encyclopedias and journals, such as the *Journal of Indo-European Studies* and *ReVision*, Dexter's work in the field of women's studies continues with her current position as the Executive Editor for the Institute of Archaeomythology. Her interest in female shamanic figures, love goddesses and sacred display is an essential element in the depiction of the archetype of the Goddess of Venus adding a much needed dimension of scholarly interest to this project.

Anthony Murphy is an avid observer and researcher of the ancient monuments in the Boyne River Valley in Ireland as well as an excellent writer who combines scientific analysis with myth and legend. His book with Richard Moore, *Island of the Setting Sun: In Search of Ireland's Ancient Astronomers* (The Liffey Press, 2008), is a masterpiece overviewing the significant archaeological sites of prehistoric Ireland along with the myths that inspired these cosmological monuments. From the beliefs and spirituality of a culture spanning eons, Murphy pieces together a study of a culture complete with the astronomically aligned monuments which commemorate their creator's idea of a cosmic unity. Murphy's expertise in archaeoastronomy and mythology is both astounding and accurate. Currently, he actively pursues a career as the creator and webmaster of *Mythical Ireland* which receives over 2,500 visitors daily, and he is the Editor of the *Dundalk Democrat*.

Barbara Carter has translated the ancient calendar which was discovered in Coligny, France. Without her insights into the astronomy of the ancients, *The Myth of the Year* and *The Goddess and the Bull* could not have been written. Carter's charts of the cycles of the sun, the moon, and Venus are an essential part of this study and are included in the appendices of the text. Her genius lies in translating the notations of the ancients which were engraved in stones of the ancient monuments in France, Britain, and Ireland and on the bronze tablet of the Coligny Calendar. Carter comes from a background as a sociologist studying at the University of Colorado and her outlook is one of ethnography combined with the expertise of astronomy; she has said she interprets the calendar by studying both the night sky and going to the culture to get the story of the people and their tribal rituals and beliefs. Her calculations and insights, and her eternal counting and re-counting of the movements of the celestial bodies visible to the ancients has endowed this study with her talents as Time-Keeper.

R. P. Hale is an interdisciplinary artist/musician, historian, and astronomer with a primary interest in the cultures, languages and astronomical accomplishments of Mesoamerica. Of Aztec ancestry himself, he carries on longstanding family traditions in printing, wood-engraving, intaglio, illustration, calligraphy and music, in his case as a concert harpsichordist, organist and hammer-dulcimer player. He was cited by the Smithsonian Institute as a premier harpsichord maker and is researching the choral and instrumental music of post-invasion Mexico and Guatemala, learning the classic Nahuatl language in the process. He is Senior Educator in astronomy, archaeo-astronomy and spectroscopy at the McAuliffe-Shepard Discovery Center in Concord, NH, teaches astronomy and is Assistant Chapel Organist at St. Paul's School, Concord, organist/music-minister for another Episcopal parish - and Craft/Folklore Coordinator at the Augusta Heritage Center in Elkins, WV. Since 1992, Hale has been a major force in debunking the 2012 "End" fraud as being highly disrespectful to his cultural background and as a Maya scholar and advocate.